FRAMING, SHEATHING and INSULATION

FRAMING, SHEATHING and INSULATION

RAYMOND P. JONES, Sr.

JOHN E. BALL

LIBRARY OF CONGRESS CATALOG CARD NUMBER: 73-1847
ISBN: 0-8273-0096-4

10 9 8 7 6 5 4 3

Printed in the United States of America
Published simultaneously in Canada
by Nelson Canada,
A Division of International Thomson Limited

DELMAR PUBLISHERS INC.
2 Computer Drive-West, Box 15-015, Albany, New York 12212

Preface

Increased demands for industrial, commercial and residential construction, combined with a shortage of skilled carpenters, have created a need for basic instructional material for industrial and vocational training programs for construction carpenters. Not only will this text meet the need of these programs, but the homeowner also will find this instructional material valuable in making repairs and performing upkeep and maintenance tasks on his home and property.

The six main occupational areas of carpentry have been determined as follows: Hand Woodworking Tools; Portable Power Tools; Concrete Form Construction; Framing, Sheathing and Insulation; Interior and Exterior Trim; and Simplified Stair Layout. Delmar Publishers has developed basic texts to cover in detail each of these areas for the carpentry student.

Each text contains those fundamental operations, procedures and data which are necessary to the achievement of certain basic skills in each of the areas of carpentry listed previously. Each step of a specific procedure that requires layout, the use of hand tools and portable power tools is fully described and is accompanied, where necessary, by line drawings.

In the development of this instructional material for *Framing, Sheathing and Insulation*, each of the units was designed to present a typical situation in the normal sequence of erecting the various framed sections that make up a completed house frame. A working knowledge of the use of the basic tools of carpentry and of concrete form construction is required of the student beginning this course of instruction.

The instructional units appear in the sequence generally followed in erecting light frame buildings. Under actual training conditions the order may be changed to meet the requirements for a locality without affecting the efficiency of the material

Grateful acknowledgment is made to the National Lumber Manufacturers Association, Technical Services Division, Washington, D.C., for supplying technical data and illustrations of current framing techniques.

Related Series

BASIC INSTRUCTION

THE USE OF HAND WOODWORKING TOOLS — A basic, comprehensive text covering measuring, layout, testing, sawing, planing, edge-cutting, and boring tools, fastening devices and abrasives.

THE USE OF PORTABLE POWER TOOLS — Covers portable electric tools such as handsaws, radial arm saws, router-shaper-planers, drills, planes, sanders. Automatic drivers, staplers, nailers, and pneumatic tools which apply to carpentry are also presented.

CONCRETE FORM CONSTRUCTION — Presents the theory and practice in layout and construction of concrete forms for carpentry construction.

FRAMING, SHEATHING AND INSULATION — Covers sill, girder and floor framing; side walls; roof and special framing practices; and insulation.

INTERIOR AND EXTERIOR TRIM — Presents the related technical information and procedures common to exterior and interior trim work for frame buildings.

RELATED BLUEPRINT READING

BLUEPRINT READING FOR BUILDING TRADES — Basic Course — Fundamental principles for reading and interpreting architectural drawings, common to all areas of the building trades.

BLUEPRINT READING FOR CARPENTERS — An advanced course with specific application of blueprint reading to carpentry work.

RELATED MATHEMATICS

BASIC MATHEMATICS SIMPLIFIED — Principles of arithmetic, algebra, geometry, and trigonometry with applications to a variety of occupations.

PRACTICAL PROBLEMS IN MATHEMATICS — FOR CARPENTERS — Selected problem material with specific applications of mathematics to carpentry work. Cross-referenced to Basic Mathematics Simplified.

Contents

Part I. SILL, GIRDER AND FLOOR FRAMING

Part II. SIDE WALLS AND SHEATHING

Part III. ROOF FRAMING

Part IV. SPECIAL FRAMING PROBLEMS

Part V. INSULATION

Unit 1   CHARACTERISTICS OF WOOD

The carpenter should have a reasonable understanding of the material he constantly uses. Under ordinary circumstances, he need not have a comprehensive knowledge of such phases of the wood industry as lumbering, milling, and the scientific aspects of tree growth. However, it is quite necessary that he understand those phases that deal with the selection, use, and care of lumber. The material in this unit will deal chiefly with the lumber used in frame construction.

WOOD STRUCTURE

Fig. 1-1 shows the cross section of a tree trunk. The bark protects the tree from injury. Just beneath the bark is a spongy layer called the cambium. This is the portion of the tree which builds new wood cells. Beneath the cambium layer is a section called sapwood. It consists of active large wood cells, which convey the sap from the root of the tree to the leaves. The rest of the trunk underneath the sapwood is called heartwood. Heartwood is formed as the tree grows and the sapwood cells mature and become inactive. Heartwood, therefore, is more desirable than sapwood for framing lumber since it is less subject to shrinking and warping as its cells are inactive and have lost their former moisture. In the center of the heartwood is the pith, which represents the growth of the tree in its first year.

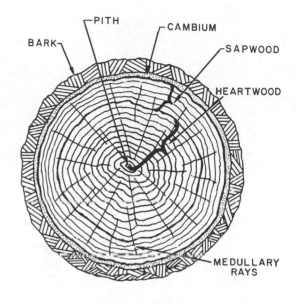

FIG. 1—1 CROSS SECTION OF A TREE

The series of concentric rings surrounding the pith are formed by the growth of the tree, one ring being added on the outside each year; hence the name, annual rings. These annual rings, as shown in the enlarged section, Fig. 1-2, are composed of fibers or long tubes running parallel to the trunk of the tree. The annual rings are crossed by fibers or cells that run from the bark of the tree to its center, conveying nourishment from the outer cambium layer to the inner part of the tree. These cells are called medullary rays.

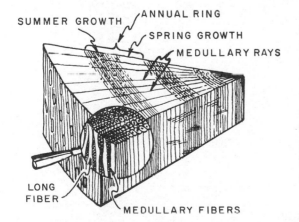

FIG. 1—2 ENLARGED SECTION

When the tubes or cells contain moisture, lumber is said to be green. Lumber is dry when the cells have collapsed and have been drained of moisture.

GRAIN AND DENSITY

Grain and density are important in determining the strength grading of lumber. Wood in which the annual growth rings are narrow is described as close grained; where the annual rings are wider apart, the wood is described as coarse grained. In softwoods, close-grained stock is stronger than coarse-grained stock.

Each annual ring consists of light springwood and dense, darker summerwood. In addition to the factor of grain, this density, or proportion of summerwood, affects the wood's strength. The higher the density, the stronger the structural lumber.

FIG. 1—3
COARSE GRAIN

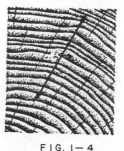
FIG. 1—4
COARSE GRAIN

FIG. 1—5
CLOSE GRAIN

METHODS OF SAWING LUMBER

The method of sawing a log has a direct bearing on its durability, quality, and ability to resist wear and to hold its shape.

Plain-sawed lumber, also called slash- or flat-sawed, is cut from the log as shown in Fig. 1-3. The log is first squared by sawing boards off the outside, leaving a rectangular section, A-B-C-D, which is then cut up as shown by the vertical lines. This is a common method of sawing framing lumber. Board E (Fig. 1-6) is shown enlarged in Fig. 1-7. Fig. 1-8 shows board F enlarged. Note that board E shows much closer annual rings as it has been cut from the heartwood portion of the log.

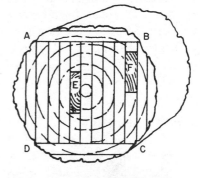
FIG. 1—6   PLAIN-SAWED LOG

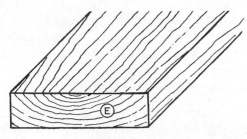
FIG. 1—7
PLAIN-SAWED INNER SECTION OF LOG

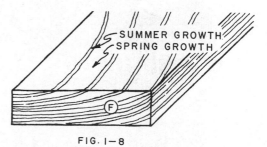
FIG. 1—8
PLAIN-SAWED OUTER EDGE OF LOG

2

The methods of quarter-sawing, Fig. 1-9, produce lumber of higher quality in some respects but cause more waste in sawing. The radial method, A, is perhaps the best because it produces lumber that will not warp as readily as that sawed by other methods. Since the annual rings are perpendicular to the surface of the board in this type of sawing, the stresses caused by the drying of the lumber will be equal in all parts of the width of the board. Furthermore, since the most shrinkage takes place in the direction parallel to the annual rings, see A, Fig. 1-10, the shrinkage in the thickness of this board is proportionally greater than in the width. Sections taken from the log as shown at A, Fig.1-10, present very short and uniform segments of the annual rings. Sections taken from B present long rings over the width of the board.

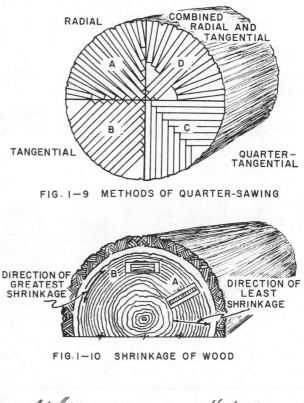

FIG. I—9 METHODS OF QUARTER-SAWING

FIG. I—IO SHRINKAGE OF WOOD

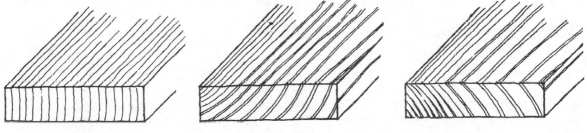

FIG. I—II
RADIAL-SAWED

FIG. I—I2
TANGENTIAL-SAWED

FIG. I—I3
QUARTER-TANGENTIAL-SAWED

The tangential cut, see section B, Fig. 1-9, is used to accomplish approximately the same results as the radial cut, but the tangential cut is more economical.

The quarter-tangential cut, see section C, Fig. 1-9, is an economical method and, except for plain sawing, is the most commonly used for framing lumber.

The radial, tangential and quarter-tangential sawed lumber, Figs. 1-11, 1-12 and 1-13, is called quarter sawed, riff sawed or edge grain. These methods are most frequently used in hardwoods to bring out the beauty of the grain, but this factor need not be considered in softwoods used in framing.

The combined radial and tangential method of sawing, see section D, Fig. 1-9, is used only for cutting large timbers because of the large amount of waste. The pieces left over can then be sawed into smaller pieces.

MOISTURE CONTENT

While a tree is living, both the cells and cell walls are filled with water. As soon as the tree is cut, the water within the cells begins to evaporate. This process continues until practically all of the "free water" has left the wood. When this stage is reached, the wood is said to be at the fiber-saturation point; that is, what water is contained is mainly in the fiber walls.

There is no change in size during this preliminary drying process and, therefore, no shrinkage during the evaporation of the "free water". Shrinkage begins when water begins to leave the cell walls themselves. These contract, becoming harder and denser, causing a general reduction in size of the piece of wood.

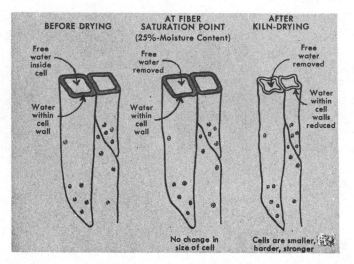

FIG. 1—14

HOW WOOD CELLS CHANGE AS WATER IS REMOVED

If a specimen is placed in an oven (kiln) which is maintained at 212°F, the temperature of boiling water, the water will evaporate and the specimen will continue to lose weight for a time. Finally, a point is reached at which the weight remains constant. This is a way of saying that all of the water in the cells and cell walls has been driven off. The piece is then said to be "kiln dry" (K.D.).

If it is now taken out of the oven and allowed to remain in the open air, it will gradually take on weight because of the absorption of moisture from the air. As in oven drying a point is reached at which the weight of the wood in contact with the air remains more or less constant. Tests show that it does not remain exactly constant, for the wood will take on and give off water as the moisture in the atmosphere increases or decreases. When the piece is in this condition, it is said to be "air dry" (A.D.).

The amount of water contained by wood in the green condition varies greatly. As a general average, at the fiber-saturation point, most woods contain from 23% to 30% water as compared with the oven-dry weight of the wood. When air-dry most woods contain from 12% to 15% moisture.

Kiln drying in great ovens is the modern, scientific system of preconditioning lumber. Under carefully controlled conditions, the moisture content of lumber for framing is reduced to an average of 19%.

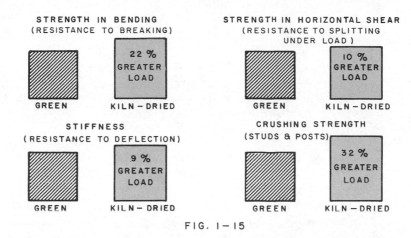

FIG. 1—15

DEFECTS AND BLEMISHES

American Lumber Standards refer to a defect as "any irregularity occurring in or on wood that may lower some of its strength, durability, or utility values", whereas, blemish is defined as "anything marring the appearance of the wood, but not classified as a defect". There are specified sizes and characteristics of defects in lumber that should not be overlooked.

A knot in lumber is caused by the growth of a branch, the inner end of which is embedded in the main stem of the tree. The location of a knot in a board may seriously affect the structural strength of the board. If the knot is solid and small, it may not do any particular harm. The knot itself is as strong as the rest of the wood, but the cross grain which develops around the knot weakens the lumber. When the lumber is being dried, checks and cracks often develop in this irregular grain. Knots can be tight or loose. Loose knots, which are formed when the wood grows around a dead branch, are apt to fall out when the log is cut into lumber. Round knots are produced when the limb is crosscut, and spike knots, when the limb is sawed lengthwise.

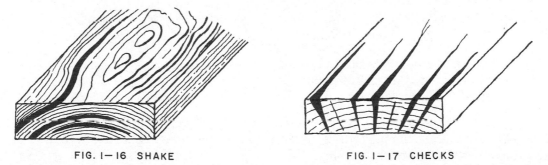

FIG. 1—16 SHAKE   FIG. 1—17 CHECKS

A shake is the separation of the wood between the annual rings lengthwise of the board, Fig. 1-16. This defect greatly weakens the board when it is subjected to a load.

A check, Fig. 1-17, is a lengthwise separation along the grain and across the annual rings. They are commonly seen on the ends of lumber and are caused by too rapid and uneven drying. They not only weaken the lumber but make it difficult to nail the board, as the nailing may cause additional splitting.

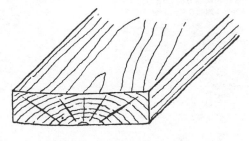

FIG. I-18 WARP

Warp, Fig. 1-18, is a bending of the lumber from a flat plane. As a board drys, it shrinks more along the long annual rings than along the short ones. Then the board will tend to curl or cup. The cupped face is on the same side as the longest annual rings.

Dry rot in lumber is caused by a fungus. The term dry rot is misleading, as it occurs only in the presence of moisture where the free circulation of air is prevented. Wet or green lumber used in a building, and so enclosed as to partially cut off air circulation, is very likely to be affected by dry rot. The fungi are sometimes found in dry wood but can draw their needed moisture long distances. Wood in the advanced stage of dry rot is shrunken, discolored, brittle, and powdery.

Some fungi do not rot the wood but only cause stain or mold which affects the wood's appearance. Wood which is blue-stained can be painted; or, if the stain is not too deep, it can be planed to restore its original appearance.

CARE OF LUMBER

The care that lumber is given after being delivered to the building site is very important. Green or partially dry lumber when not properly piled will twist and warp in drying and will retain this twisted and warped shape. The lumber should be protected from the rain and should be piled in such a manner that air can freely circulate through the pile. If the lumber is piled tightly together and is allowed to get wet, an infection may start and continue in the lumber after it has been placed in the building.

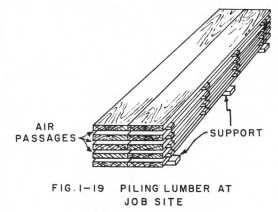

AIR PASSAGES

SUPPORT

FIG. I-19   PILING LUMBER AT JOB SITE

Lumber that has been used for concrete forms, scaffolds, and staging should be properly cleaned and inspected for defects caused by rough handling before it is used for permanent parts of the building.

SURFACING OF FRAMING LUMBER

Framing lumber is classified as rough lumber and dressed lumber. Rough lumber is lumber as it comes from the saw. Dressed lumber is lumber that has been surfaced by running it through the planer. It may be surfaced on one side (S1S), two sides (S2S), or a combination of sides and edges (S4S). Most framing lumber is S4S.

STANDARD SIZES OF FRAMING LUMBER

Dimensions of milled lumber are always scant, since the wood must be seasoned and smoothed after it is measured and cut. For instance, a board nominally 1" thick is actually 3/4", while a 2" x 4" measures about 1/2" less each way, or 1 1/2" × 3 1/2". Boards nominally 8" or more wide lose as much as 3/4" in seasoning and milling.

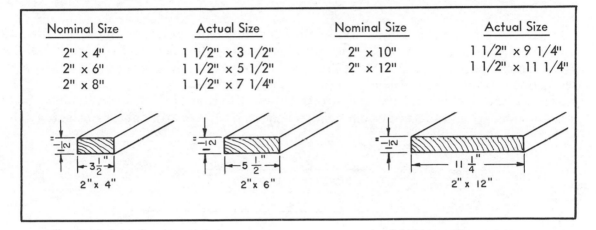

Nominal Size	Actual Size	Nominal Size	Actual Size
2" x 4"	1 1/2" x 3 1/2"	2" x 10"	1 1/2" x 9 1/4"
2" x 6"	1 1/2" x 5 1/2"	2" x 12"	1 1/2" x 11 1/4"
2" x 8"	1 1/2" x 7 1/4"		

KINDS OF LUMBER USED FOR FRAMING

There are four varieties of wood generally used for framing lumber: southern yellow pine, Douglas fir, spruce and hemlock. Southern yellow pine is found in the southern states from Virginia to Texas. There are two varieties: the long leaf or heavy close-ringed species, and the shortleaf which is lighter in weight and softer in texture. The heartwood of southern pine is highly resistant to decay and can be easily treated against attacks by termites and dry rot. It has a straight grain and is an excellent wood for structural members. The less dense or shortleaf pine is most commonly used for light building construction and for trim. It is lighter, works more easily, and holds paint better than the longleaf variety.

Douglas Fir is found largely in the states of Oregon and Washington. It is very strong and has a straight, stringy, and tough grain. It is used where considerable strength is needed in a framework, as well as for ordinary structural members.

Spruce may be classified as white, red, black, and western. The western spruce is strong and light in weight, with a straight, even grain. Red spruce is nearly white, with a reddish tinge. It is light in weight and has an even grain with a fine texture. White spruce has a coarser grain than the others. Spruce is frequently used for scaffolds, studs, and lath.

Hemlock may be divided into eastern hemlock and west coast hemlock. Eastern hemlock is light in weight, has a straight grain, and is quite splintery. The heartwood is brown, while the sapwood is a lighter brown. It is quite brittle and has small knots, but it is cut extra thick to offset its structural weakness. West coast hemlock is hard, strong and stiff. It is extensively used for framing as well as for flooring, paneling, and interior trim.

GRADES OF LUMBER USED FOR FRAMING

Dimension lumber is judged primarily on the basis of strength. There are two grade groups, common dimension and structural dimension. Common dimension lumber is 2″ or more in thickness and from 2″ to 12″ wide. The lengths are in multiples of 2 feet from 6′ to 20′ long. The structural dimension lumber is always 4″ or more wide. Any lumber more than 5″ in width and thickness is called timber.

The numbering of dimension lumber indicates the relative amount of defects allowed. No. 1 dimension lumber permits tight knots, pitch pockets, and other defects that do not impair strength. No. 2 dimension lumber requires slightly less strength than No. 1. No. 3 dimension allows large knots, knot holes, decay, and checks. The graph and table which follow indicate the classification of dimension lumber, from best to poorest, and the minimum grades required for certain species of softwoods.

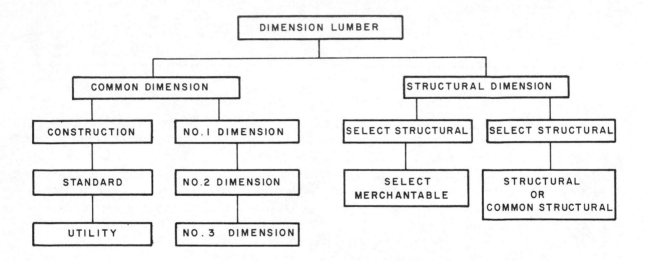

Species	Minimum Grade	Species	Minimum Grade
Cypress	No. 1 com. dimension.	Pine, Red or Norway	No. 1 dimension.
Douglas Fir, WCLA	Standard Grade.	Pine, Southern Yellow	No. 2.
Douglas Fir, WPA	No. 1 dimension.	Pine, Western White	No. 1 dimension.
Fir, Balsam	No. 1.	Red Cedar, Western, WCLA	Construction Grade.
Fir, White, WCLA	Construction Grade.	Red Cedar, Western, WPA	No. 1 dimension.
Fir, White, WPA	No. 1 dimension.	Redwood, California	Sap common.
Hemlock, Eastern	No. 1 com. dimension.	Spruce, Eastern	No. 1 dimension.
Hemlock, Western	Standard Grade.	Spruce, Engelmann	No. 1 dimension.
Larch, Western	No. 2 dimension.	Spruce, Sitka	Construction Grade.
Pine, Eastern White	No. 1 dimension.		

REVIEW PROBLEMS   Unit 1

1. Why is heartwood more desirable than sapwood for framing lumber?

2. How do you distinguish between close-grained and coarse-grained lumber?

3. Explain how grain and density affect strength of lumber.

4. Which method of sawing produces a higher quality lumber? Why?

5. In what direction does most shrinkage take place in lumber?

6. How do you define fiber-saturation point?

7. At the fiber-saturation point, is there any change in size of the lumber?

8. List some advantages of air-dried wood over green wood.

9. List advantages of kiln drying over air drying.

10. Fill in the actual sizes of the following standard lumber sizes.

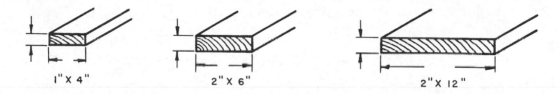

1" X 4"   2" X 6"   2" X 12"

11. On the illustration shown, identify the types of cuts and explain why the shrinkage and distortion would appear as illustrated.

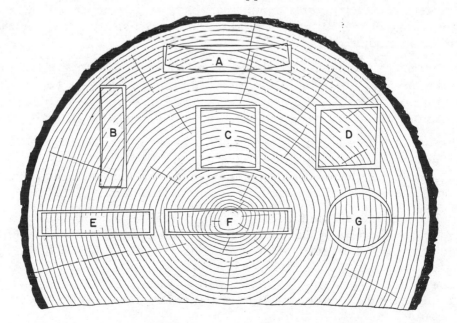

Unit 2   TYPES OF HOUSE FRAMING

There is a difference of opinion among carpenters as to the best type of frame construction. This is possibly due to the fact that, throughout the country, several different systems of framing are used. In most cases it is attributable to economic reasons rather than to differences in structural data obtainable. It is the purpose of this unit to acquaint the learner with the approved methods of framing. The detailed assembly of the complete frame and such sections of the frame as girders, sills, joists, and subfloors that are common to all types of framing will be considered in separate units.

CLASSES OF STRUCTURES

Buildings constructed entirely of wood above the foundations may be classified as follows: early braced frame, modern braced frame, western or platform frame, balloon frame, and plank-and-beam frame. The latter three framing methods have come about because of the rapid development of woodworking machinery, the improvements in the manufacture of building materials, and the competition among contractors for low cost and high production.

FUNDAMENTALS OF FRAME CONSTRUCTION

During the lifetime of a building, the lumber of which it is constructed undergoes many changes of temperature and humidity. Even well-seasoned lumber will dry out further under the artificial heat of a house in winter. As the moisture content is reduced, the lumber shrinks somewhat. The amount of shrinkage depends on the moisture content of the structural member at the time it is set in place, the temperature and humidity of the building, and also the compression caused by the load on the structural member.

The shrinkage occurs primarily across the board (3%-6%), very little taking place lengthwise (1/10 of 1%). Therefore, the more horizontal bearing members there are in a house frame, the more shrinkage there will be in the frame. This fact is of utmost importance and should guide the selection of framing methods.

In the early braced-frame construction most of the bearing members were vertical. This type of construction presented very little shrinkage in the frame of the building. Some of the modern systems of framing have retained this principle as far as possible, while others have, to some extent, disregarded it.

The side and end walls of a frame are generally braced by inserting diagonal (45°) braces at the corners of the building or by running the sheathing diagonally (45°) to the studs in opposite directions from each corner. The walls should be tied together by adequate joists, which should be braced to each other by bridging at intervals not exceeding 10 feet. Openings in walls, partitions and floors should be provided with headers strong enough to carry the load of the studs or joists that have been cut to make the opening.

If these fundamental principles of frame construction are given major consideration, any type or combination of types of frames may be used.

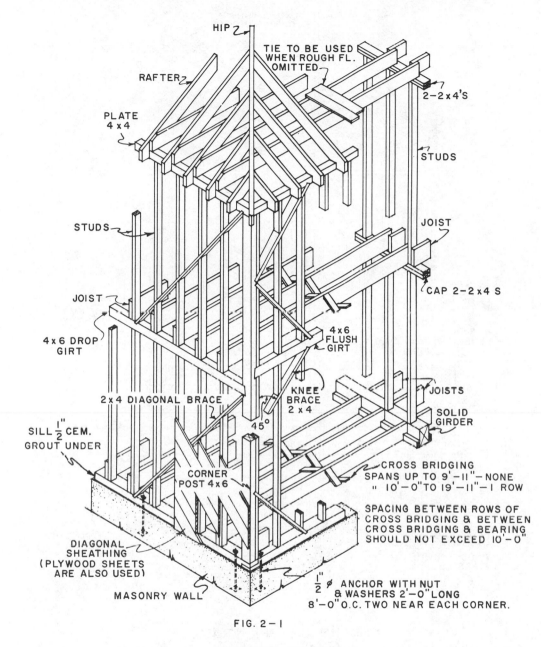

FIG. 2-1

BRACED FRAME

In the early braced frame, each post and beam of the framework was mortised and tenoned together, and the angles formed by these posts or beams were held rigidly by diagonal braces. These braces, in turn, were held by wooden pins or dowels through the joints. This type of frame construction did not rely on the sheathing for rigidity; in fact, the sheathing or planking could run vertically up the sides of the structure. Barn construction was typical of this type of building. The present conventional braced framing has replaced this type of construction and is used to a large extent where sheathing is not entirely relied upon for the rigid support of the building.

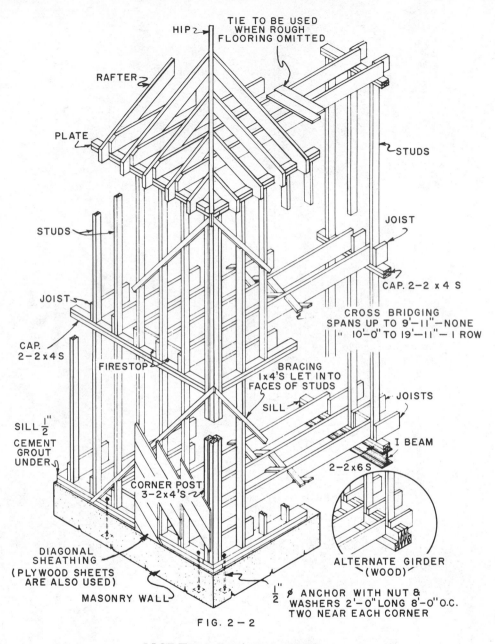

FIG. 2 — 2

MODERN BRACED FRAME

In the modern braced frame, the studs on the side walls and the center bearing partitions are cut to exactly the same length. The side wall studs are attached to the sill plate, and the partition studs are attached to the girder and plate. This construction permits uniform shrinkage in the outside and inside walls if a steel girder is used. Diagonal braces are let into the studs at the corners of the building to provide rigidity.

The joists are lapped and spiked to the side wall studs and are continuous between the two opposite side walls, or are side lapped and spiked at the center partition. In either case, these joists form a tie for the two side walls. The joists are bridged and are supported by their full width at the sill and girder. The subflooring is laid diagonally to the floor joists to provide rigidity to the building as well as to the floors.

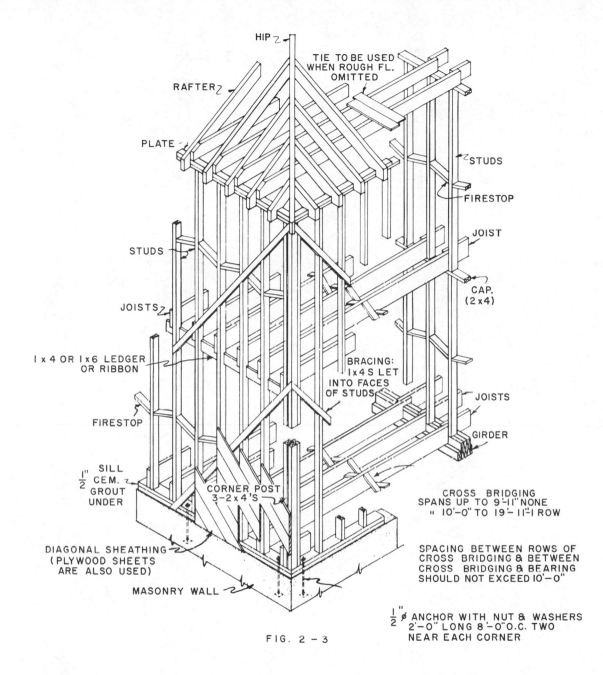

FIG. 2 − 3

BALLOON FRAME

Fig. 2-3 shows the balloon frame. The principal characteristics of balloon framing are that both the wall studs and joists rest on the anchored sill, plus the fact that the studs extend in one piece from the foundation to the roof. The second-floor joists are supported by a ribbon let into the inside face of the studs, and the joists are also fastened to the studs to support and tie the structure together. This system of framing replaced the modern braced frame until platform or western framing became popular because of the demand for one-floor homes and split-level homes.

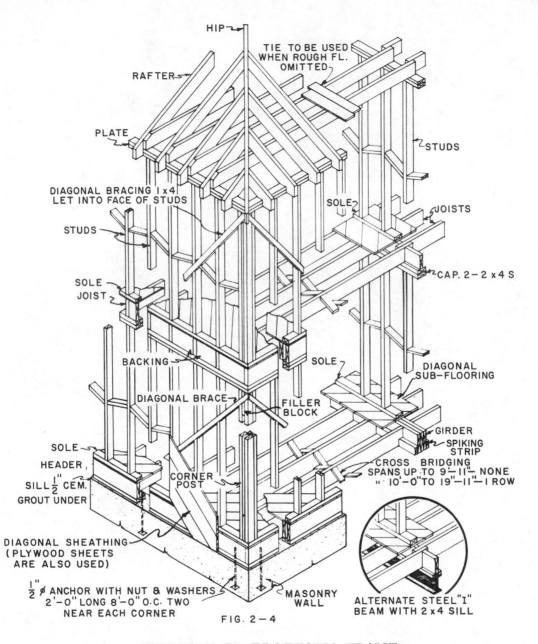

HIP

TIE TO BE USED
WHEN ROUGH FL.
OMITTED

RAFTER

PLATE

STUDS

DIAGONAL BRACING I x 4
LET INTO FACE OF STUDS

SOLE

JOISTS

STUDS

SOLE
JOIST

CAP. 2—2 x 4 S

BACKING

SOLE

DIAGONAL BRACE

DIAGONAL
SUB-FLOORING

FILLER
BLOCK

GIRDER

SOLE

SPIKING
STRIP

HEADER

CORNER
POST

CROSS BRIDGING
SPANS UP TO 9'—11"— NONE
" 10'—0"TO 19'—11"—I ROW

SILL 1½" CEM.
GROUT UNDER

DIAGONAL SHEATHING
(PLYWOOD SHEETS
ARE ALSO USED)

½" ⌀ ANCHOR WITH NUT & WASHERS
2'-0" LONG 8'-0" O.C. TWO
NEAR EACH CORNER

MASONRY
WALL

ALTERNATE STEEL "I"
BEAM WITH 2 x 4 SILL

FIG. 2 — 4

WESTERN OR PLATFORM FRAME

Fig. 2-4 shows the western type of frame. Each story of the building is built as a separate unit, the subfloor being laid before the side walls are raised. The exterior walls may be assembled as full wall sections on the subfloor. The bottom and top plates are nailed to the precut studs, and the window and door headers are fastened in place. The notches are cut and the diagonal cut-in braces are nailed in place. Then the section is raised, the bottom plate fastened to the platform, and the top plate doubled to join the sections together and hold the wall straight. This provides safer working conditions and could be used in the other types of framing. This type of framing disregards the principle of shrinkage but permits equal shrinkage at the side and center walls.

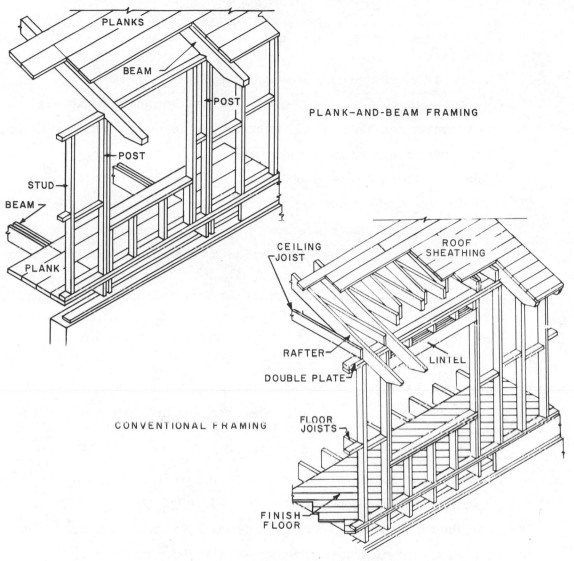

FIG. 2—5   PLANK-AND-BEAM FRAMING COMPARED TO CONVENTIONAL FRAMING

PLANK-AND-BEAM FRAMING

The plank-and-beam frame is shown in A, Fig. 2-5. Plank-and-beam framing uses a plank subfloor or roof decking with supporting beams, which may be placed up to seven feet apart. This framing method uses a few larger members to replace many small pieces necessary in conventional framing. It may be used as a floor system and/or a roof system in buildings having either balloon or platform framing. When it is used as a roof system, the exposed ceiling material of the room below should be selected for good appearance. Lath and plaster may be omitted by placing the insulation on top of the planks of a roof system and finishing the underside, or by placing a decorative exposed insulation on the underside. When used as a floor system, the finish flooring should be laid at right angles to the planks of the subfloor. This type of framing can be constructed by using either solid or built-up posts and beams.

REVIEW PROBLEMS   Unit 2

1. Why is there a difference of opinion as to the best type of house frame construction?

2. List five types of house frame construction.

3. State three reasons why the methods of house framing are changing.

4. Does lumber, in a structure, undergo any changes during its lifetime?

5. What makes seasoned lumber shrink?

6. Explain the four reasons which cause a structural member to shrink.

7. State the percentage of shrinkage that takes place across the grain of lumber.

8. What percentage of shrinkage takes place lengthwise in lumber?

9. Is it advisable to use many horizontal bearing members in a house frame? Why?

10. What fundamental fact should be of utmost importance in selecting a method of house framing?

11. Describe two ways in which the sidewalls of a house may be braced.

12. How far apart should bridging be placed?

13. How were the posts and beams of the early braced frame held together?

14. Is the factor of uniform shrinkage considered in the modern braced frame? Explain.

15. What is the advantage of diagonal flooring?

16. Explain the principal characteristics of balloon framing.

17. How are the second-floor joists of the balloon frame supported?

18. Describe the difference between balloon and platform framing.

19. How does platform framing provide safer working conditions?

20. What is meant by equal shrinkage in house construction?

21. How far apart may beams be placed in plank-and-beam construction?

22. Can plank-and-beam construction be combined with other types of framing? Explain.

23. Describe two methods of finishing the ceilings of a plank-and-beam-framed house.

Unit 3   WOOD-SILL CONSTRUCTION

That part of the side walls of a house that rests horizontally on the foundation wall is called the sill. The member which actually contacts the foundation wall is called the sill plate. For certain types of sills a sill header is a part of the sill construction. This member is attached to the sill in a horizontal position on edge as shown in Fig. 3-5. The various types of sills are named according to the location of the sill header on the sill plate. The important factors to be considered in the design of sills are (1) that a solid and rigid sill plate be provided for the support of the joists and side walls; (2) that the sill be so constructed that it provides a means of tying the framework and masonry wall together with an immovable airtight joint; and (3) that the type of sill construction be similar to that of the girder so the shrinkage at these bearing points will be equal.

SOLID TYPE OF SILL

The old-style solid sill is a sill plate, usually $4'' \times 6''$, bolted firmly to the foundation wall, upon which the joists and side wall studs rest. The joint at the corner is half-lapped together. This type of sill was used in the early type of balloon frame and has since been modified.

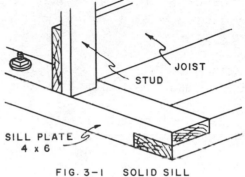

FIG. 3-1   SOLID SILL

MODERN TYPE OF SOLID SILL

The modern type of solid sill which may be used in balloon frame is built with modern sized lumber. The sill plate should be set true and level and provide full bearing on the foundation wall. The corner joints should be butted together and strapped. The side wall studs rest directly on the sill plate. Blocking is inserted between and level with the top edge of the joists to carry the ends of the subfloor. This type of sill may be built up by using a double sill plate. Usually a $2'' \times 10''$ lower sill plate and a $2'' \times 4''$ upper sill plate are installed. The double sill plate may be straightened more easily than the single plate and provides for the lapping of one plate over the other at joints and corners of the building.

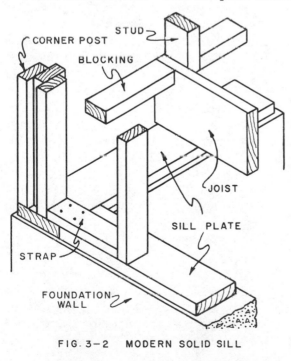

FIG. 3-2   MODERN SOLID SILL

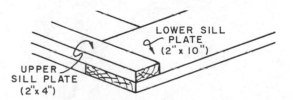

FIG. 3-3   LAP OF DOUBLE SILL PLATE

17

"T" TYPE OF SILL

The "T" sill is a built-up sill which uses two members. The sill plate rests directly on the foundation wall and the sill header is centered on the sill plate. The header provides solid sill construction and a firm base upon which to nail the subfloor. This header may be cut in between the joists or may be spiked to the ends of the joists. The method used depends upon the width of the sill plate. The joists should have a bearing seat on the sill plate of at least 4". Thus, if the sill header is spiked against the ends of the joists, as shown in Fig. 3-4, the sill plate would have to be 2" × 10".

If the sill header is cut in between the joists in the form of blocks of wood from joist to joist, the sill plate can be 2" × 8".

FIG. 3-4   "T" SILL

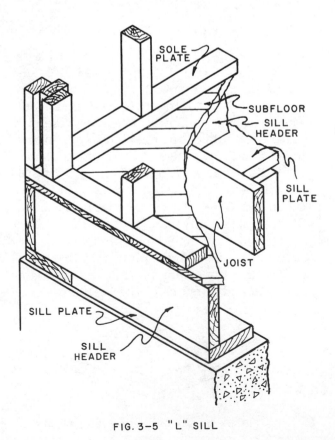

FIG. 3-5 "L" SILL

"L" TYPE OF SILL

The "L" sill is the type of sill construction generally used in the platform type of frame. This sill provides a solid bearing upon which to nail the subfloor. Notice that the subfloor runs to the outside of the sill header and must be laid before the side wall is raised. The disadvantage of this type of sill is that the side wall studs rest on horizontal grain of considerable thickness. This thickness is equal to that of the sole plate, subfloor joist, and, sill plate. This disadvantage may be overcome by using the same combination of members at the bearing partitions or by using trussed roof construction. The "L" sill construction is erroneously called a "Box Sill"; but, inasmuch as the subfloor separates the sill header from the sole plate, it can not be considered as a single unit.

METHODS OF FASTENING SILLS TO FOUNDATION WALLS

Wood sills are fastened to masonry walls by 1/2" bolts with a 2" washer. The bolt should be embedded 15" or more in the wall. The maximum spacing of the bolts should be 8' o.c., with not less than two bolts in each sill piece. The end bolts should be not more than 1'-0" from the ends of each piece.

When sill plates are fastened to concrete walls, the 1/2" bolts should be embedded not less than 6" in the wall. The spacing and location of the bolts should be the same as for masonry walls.

Hardened steel studs driven by powder-actuated tools may be used to fasten the sills to foundation walls except in earthquake-designated areas. The studs must not be more than 4" apart, and there should be at least two in each sill piece.

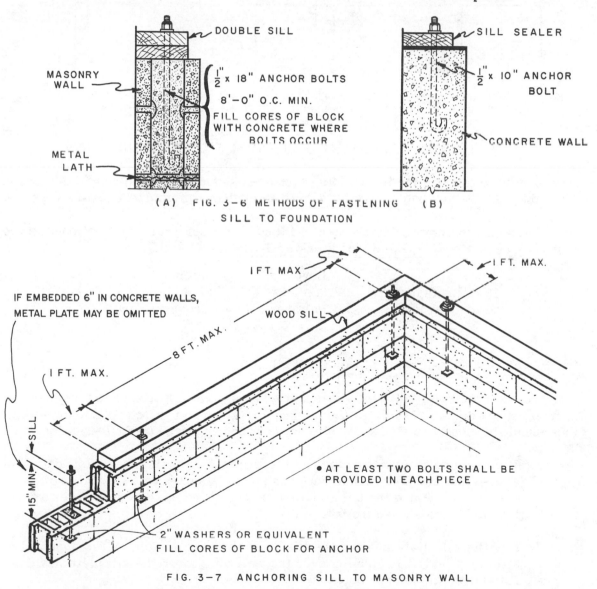

FIG. 3-7   ANCHORING SILL TO MASONRY WALL

PLACEMENT OF SILLS ON PIERS

When a load is placed on a pier by the joists and sill, the load is concentrated, and the line of force is approximately in the center of the sill plate. Therefore, the sill plate should be placed as near the center of the pier as possible to avoid tipping the pier as shown.

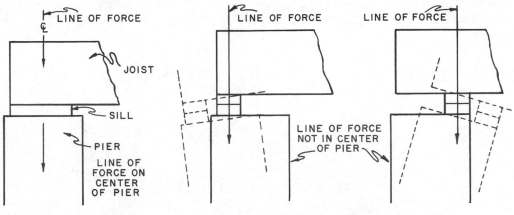

FIG. 3—8   CORRECT AND INCORRECT POSITIONS OF A SILL PLATE

TOOLS AND EQUIPMENT FOR WOOD SILL CONSTRUCTION

The following are the basic tools which are used in the framing and installation of wood sills. Power tools may be substituted wherever and whenever available.

Information concerning the use of these and other tools may be found in the first book of this series, Hand Woodworking Tools.

Rule	Level	Wrench
Framing Square	Steel Tape	Straightedge
Brace and Bit	Hammer	Crosscut Saw

How to Frame and Install the Sill on the Wall

The accuracy with which the sill is framed is of great importance to the successful framing of the entire building. The dimensions given on the floor plans and elevations should be closely studied so the sill may be properly placed on the foundation wall.

1. Place the sill plates along the top of the four walls of the foundation in approximately their permanent positions. Use 14′ or 16′ lengths of stock if possible. Place the sill so no joints occur over an opening in the foundation wall, and square the butt joints.

2. Lay the sill plate against the inside of the anchor bolts. The ends of the sills may be left projecting over the wall until after the sill plate has been fitted over the anchor bolts.

3. Square lines from each anchor bolt across the sill plate.

4. Measure the correct distance along each line from the outside edge of the sill to locate the holes for the anchor bolts.

   NOTE: In most cases, the outer edge of the sill should be in from the outside edge of the foundation wall so the sheathing will come flush with the foundation wall.

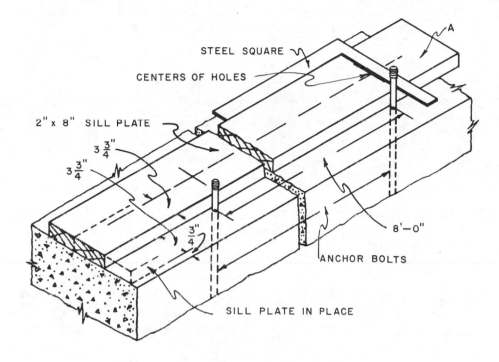

FIG 3-9 METHOD OF MARKING SILL FOR ANCHOR BOLT HOLES

5. Mark these points and bore or drill the holes with a bit 3/8″ larger than the diameter of the bolts. This size hole will allow for straightening the sill and will make it easy to place the sill plate over the bolts.

   NOTE: If termite shields are to be used, it is well to insert them before the sill plate is placed over the bolts. If the foundation wall is level and true, a sill sealer may be used and should be placed at this time.

6. Place the sills over the anchor bolts.

7. Cut the sill plates at the corners of the building so they butt together closely.

8. Place washers and nuts on the anchor bolts and draw the nuts down to bring the sill plate down temporarily to the masonry wall.

9. Square the plate at the corners, using the 6-8-10 rule, and brace the sill in this position.

10. If the foundation wall is not level and true, level and straighten the sill over the bearing walls by using a straightedge and level. If shims are needed, place them every 4' between the bottom of the sill plate and the top of the foundation.

    NOTE: Some carpenters prefer to do this after the floor joists have been placed so the weight of the joists will hold the sill plate down against the wall.

11. After the top of the sill plate is level and straight and the outside edge is straight with the face of the wall, tighten the nuts on the anchor bolts, but not so much that they will pull the sill out of alignment.

12. Place the header and toenail it to the sill with 10d nails, 16" on centers, so it stands squarely with the face of the wall.

    NOTE: The sill plate should be set in a full bed of Portland cement mortar when necessary to obtain full bearing.

How to Frame and Install the Sill on Piers

NOTE: With a pier foundation, the sills are built up like girders to support the load of the joists over the span between the piers. Anchors, flat bars projecting from the top of the foundation wall, are incorporated in the built-up girder as it is being built.

1. Align and nail together the two outside members of the sill, using 20d nails, two near each end of each piece, other nails to be staggered 32" apart. Be sure the tops of the members are even and the crowned side is up.

2. Place the sill on the piers and against the iron strap anchors in the piers. Do not put any joints of the built-up sill over the span between the piers.

3. Fasten the anchors to the side of the sill, using 20d nails.

4. Place the remaining member of the sill against the opposite side of the anchor and mark the location and size of the anchor on this member.

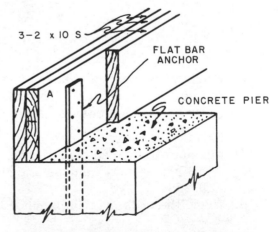

3-2 x 10 S

FLAT BAR ANCHOR

A

CONCRETE PIER

FIG. 3-10 ANCHORING A BUILT-UP SILL TO A PIER

5. Notch out this piece to receive the flat anchor bars so that it will fit up tightly against the other two members.

6. Fasten the third piece to the other pieces, using 20d nails, two near each end of each piece, other nails to be staggered 32" apart.

7. Level the sill from pier to pier and grout with Portland cement mortar.

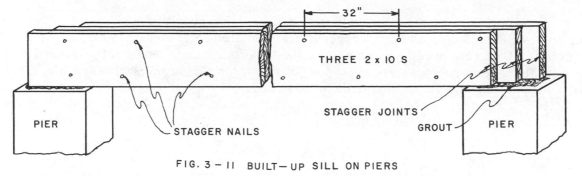

FIG. 3 – 11 BUILT–UP SILL ON PIERS

REVIEW PROBLEMS Unit 3

1. Describe a house sill.

2. Name and locate the various members of a house sill.

3. How are the different types of sills named?

4. List the three important factors to be considered in the design of sills.

5. Describe the old-style solid sill as compared with the modern-style solid sill.

6. Why is a double sill plate sometimes used?

7. When using a "T" sill, how much bearing should the joists have?

8. How can the "T" sill be constructed if the sill plate is only 8" wide?

9. Which type of framing uses the "L" sill?

10. What is the disadvantage of the "L" sill and how can it be overcome?

11. How should a sill be placed with relation to the centerline of a pier?

12. When installing a sill plate, how can the anchor bolts be located?

13. How far in from the face of the foundation wall is the sill plate usually placed?

14. How much larger than the diameter of the anchor bolt should the holes in the plate be drilled? Why?

15. When is it permissible to use a sill sealer?

16. How can the corners of sill plates be squared?

17. When necessary to level a sill plate, how far apart should shims be placed?

18. What size nails and how far apart should they be placed when fastening the sill header to the sill plate?

19. If the top of the foundation wall is not straight and true, how should the sill plate be set to obtain a full bearing?

20. Sills for piers are built up like girders. What size nails are used, and how should they be placed?

Unit 4   GIRDERS

A girder is a large beam (horizontal member) that supports other smaller beams or joists. It may be made up of several beams nailed together, or it may be of solid wood, steel, or reinforced concrete, or a combination of these materials. A girder is generally used to support the ends of the joists over a long span, thus taking the place of a supporting partition. The girders carry a very large proportion of the weight of a building. They must be well designed, rigid, and properly supported, both at the foundation walls and on the columns.

TYPES OF GIRDERS

The simplest and most common form of wood girder consists of a single piece. It may be as small as $2'' \times 4''$ for light ceiling joists or rafters in residences and as large as $12'' \times 24''$ where heavy loads are to be supported over long spans.

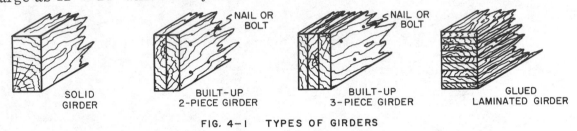

FIG. 4-1   TYPES OF GIRDERS

The built-up girder is very commonly used in house construction. It is generally made of three members fastened together with the joists resting on top of the girder.

When built-up girders are continuous over three or more supports, joints in the girders may be located between 1/6 and 1/4 the span length from the intermediate support. No two adjoining members nor more than one third the total number are to be joined on the same side of the support. The girders should be securely anchored to masonry piers, nailed to wood posts, or bolted to steel columns.

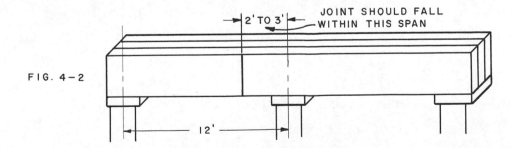

FIG. 4-2

Wood joists may be framed into the sides of wood girders in several different ways. One method is to rest the joists upon a ledger board and fasten them to the girder with at least three 10d nails. The joists should not be notched more than 1/4 of their depth. The minimum size of the ledger should be a piece of 2×2, with three 16d nails at each joist. The nails to be placed approximately three inches on centers in the 2×2. Another method is to use two pieces of 3×3 angle six inches long made of at least 18-gage metal with six nails in each angle. Still another method is to use a commercial type of steel joist hanger made of $1/4 \times 1/2$ strap iron.

24

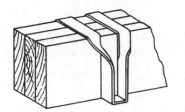

FIG. 4—3   GIRDER WITH JOIST
HANGER

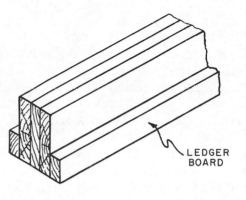

LEDGER
BOARD

FIG. 4—4   GIRDER WITH LEDGER BOARD

Fig. 4-3 shows a girder over which joist hangers have been placed to carry the joists. This type is used where there is little headroom, or where the joists carry an extremely heavy load.

Fig. 4-4 shows a girder with a ledger board, upon which the joists ride, spiked along each side. This construction is used to conserve headroom.

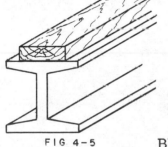

Steel wide-flange (WF) beams may be used instead of a wooden girder, Fig. 4-5. Two-inch stock is sometimes used on top of the steel beam for the joists to rest upon, to provide a surface to which to nail the joists, and to provide the same amount of horizontal-grain wood to support the load as there is in the outside wall.

FIG 4-5   BEARING SEATS OF GIRDERS

The ends of a wood girder should not be embedded in a masonry wall in such a way as to prevent free circulation of air around the ends. If this precaution is not taken, dry rot may occur at this point. It is good procedure to rest the girder on a bearing plate of iron that is set in the wall. This bearing plate should be at least 2″ wider on each side than the girder. Bearing plates range from 3/8″ to 1″ in thickness and help distribute the load of the girder over a larger area of the wall.   They are particularly important where built-up beams are supported by masonry walls. Wood shims should not be used under girders.

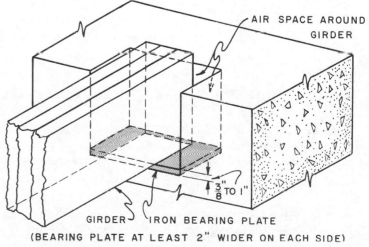

AIR SPACE AROUND
GIRDER

3/8″ TO 1″

FIG. 4—6   WOOD GIRDER
BEARING ON MASONRY
WALL

GIRDER   IRON BEARING PLATE
(BEARING PLATE AT LEAST 2″ WIDER ON EACH SIDE)

SPACING OF GIRDERS

The length and the depth of the joists and the location of the bearing partitions of the floor above must be considered in determining the spacing of girders. It is good procedure to keep the distance between girders 15 feet or less. Thus, in a span of 25 feet, one girder is sufficient if it is placed halfway between each of the other supports. If the distance is 35 feet, two girders equally spaced are necessary. Girders should be placed below bearing partitions. This restriction sometimes changes the spacing of girders between side wall supports. If there are two bearing partitions, there should be two girders regardless of the span, Fig. 4-7.

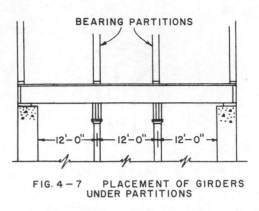

FIG. 4 – 7   PLACEMENT OF GIRDERS UNDER PARTITIONS

A girder under joists having no bearing partitions above is shown in Fig. 4-8. Fig. 4-9 shows a bearing partition placed midway between two side walls. The girder is directly beneath this partition. The bearing partition in Fig. 4-10 is 8 feet from one side wall and 15 feet from the other. Again, the girder is located directly underneath the partition.

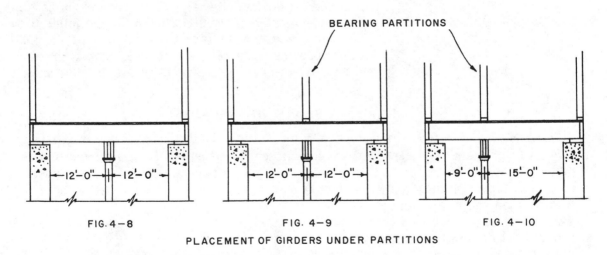

FIG. 4 – 8   FIG. 4 – 9   FIG. 4 – 10

PLACEMENT OF GIRDERS UNDER PARTITIONS

EFFECT OF DIMENSIONS ON STRENGTH OF GIRDERS

The girder must be large enough to support whatever load may be imposed on it, but not so large that it is wasteful. Before determining the size of a girder or beam the carpenter should understand three important relationships:

1. The effect of length of a girder on its strength.
2. The effect of width of a girder on its strength.
3. The effect of depth of a girder on its strength.

Length

If a plank supported at the ends carries an evenly distributed load throughout its entire length, it will bend to some extent. A plank twice as long, with the same load per foot of length, will bend much more and may break. If the length is doubled, the safe load will be reduced not to one-half, as might be expected, but to one-quarter. However, for a single concentrated load at the center, doubling the length decreases the safe load by only one-half.

The greater the unsupported length of the girder, the stronger the girder must be. The strength may be increased by using a stronger material or by using a larger beam. The beam may be enlarged by increasing the width or depth or both.

Width

Doubling the width of a girder doubles its strength. This double-width girder will have a load-carrying capacity equal to two single-width girders placed side by side.

Depth

Doubling the depth of a girder increases its carrying capacity four times. A beam 3″ wide and 12″ deep will carry four times as much as one 3″ wide and 6″ deep. Therefore, to secure additional strength, it is more economical to increase the depth of a beam than the width. However, it is well to avoid increasing the girder depth to much more than 10 inches, since a deeper girder will cut down the headroom in the basement. Instead, the girder could be made wider, additional supports could be put in to reduce the span, or a stronger material could be used for the girder.

PROPORTION OF THE TOTAL FLOOR WEIGHT CARRIED BY A GIRDER

A single girder running through a building (whether or not it be at the center of the building), and supporting the inner ends of the floor joists, will carry half the weight of every joist resting upon it. Therefore, the girder carries half the weight of the floor, while the two foundation walls that support the outer ends of the joists divide the other half of the weight between them.

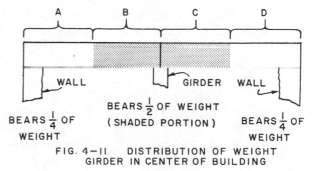

FIG. 4—11   DISTRIBUTION OF WEIGHT
GIRDER IN CENTER OF BUILDING

In Fig. 4-11, the proportion of the total weight carried by one wall is represented by A. The proportion of the total weight carried by the girder is represented by B plus C. The proportion of the total weight carried by the second wall is represented by D.

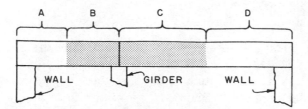

FIG. 4−12   DISTRIBUTION OF WEIGHT.
GIRDER OFF CENTER OF BUILDING

If the girder is not in the center of the building, the weight will be distributed as shown in Fig. 4-12. The proportion of the total weight carried by one wall is represented by A. The proportion of the total weight carried by the girder is represented by B plus C, while the proportion of the total weight carried by the second wall is represented by D. In general, a girder will carry the weight of the floor on each side to the mid-point of the joists that rest upon it.

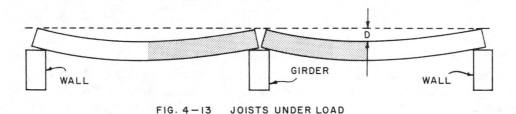

FIG. 4−13   JOISTS UNDER LOAD

It has been assumed that the joists are butted or lapped over the girder. Loaded joists have a tendency to sag between supports (Fig. 4-13) and when they are butted or lapped, there is no resistance to this bending over the girder. An exaggerated amount of this sag is shown by D in Fig. 4-13.

If the joists are continuous, however, under load they will tend to assume the shape shown in Fig. 4-14. Being in one piece, they resist bending over the center support, thereby forcing the girder to carry a larger proportion of the load than if the joists were cut. A girder at the mid-point of continuous joists will take five-eighths instead of one-half the floor load.

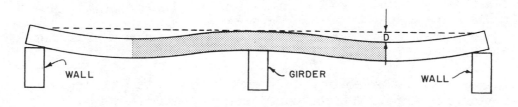

FIG. 4−14   CONTINUOUS JOIST UNDER LOAD

DETERMINING THE SIZE OF A GIRDER

The following steps are necessary in determining the size of a girder and will be explained by an example:

1. Find the distance between girder supports.
2. Find the total floor load per square foot carried by the joists and bearing partitions to the girder.
3. Find the load per linear foot on the girder.
4. Find the total load on the girder.
5. Select the material for the girder.
6. Find the proper size of the girder.

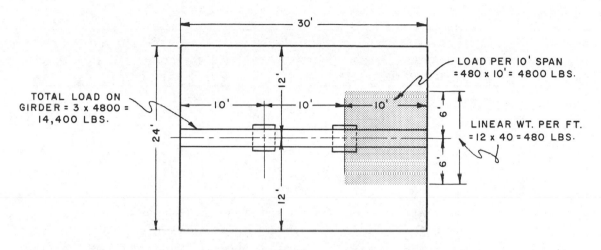

FIG. 4—15   HOW TO DETERMINE SIZE OF GIRDER TO BE USED

Before it is possible to figure the girder size, the length of the girder between supports must be determined. This depends on the spacing of the supporting posts. These posts must be spaced according to some suitable division of the total length of the girder. For the purpose of illustration, assume that the total girder length is 30 ft. As posts carry a larger part of the weight, and since the use of short spans will reduce the size of the girders, it will be desirable to use two posts. Therefore, the span will be 10 feet, Fig. 4-15.

Assume that the building is to be 24' wide, the girder being placed 12' from each side wall support, and the joists cut or lapped over the girder. The total load per square foot on the floor is assumed to be 40 pounds.

The load per linear foot on the girder may be found by adding together the half-widths of the joist spans (6 ft. + 6 ft. = 12 ft.) and multiplying this sum by the load per square foot (12 ft. × 40 lbs. = 480 lbs. per linear foot of girder). Since the supporting columns under the girder are 10 feet apart, the total load on the girder of 10 ft. span would be 10 ft. × 480 lbs. or 4800 lbs.

If the material of the girder is to be Douglas fir or southern pine, the chart on the following page may be used.

In the column under Span in Feet marked 10, follow the figures down until the figure 4800 (the total load in this example) or the next larger figure, which happens to be 6406, is found. The horizontal column is followed to the left where, under the column headed Solid Girder Size, the figures 8 × 8 are found. This indicates that a solid beam 8 × 8 will be needed.

If the girder is to be built up of four 2 × 8 s, the figure 6406 should be multiplied by .867 as indicated at the bottom of the chart. Thus, the carrying capacity of the built-up beam would be 6406 × .867 = 5554 lbs.

Following the same procedure in reading the chart, check to see if a girder other than an 8 × 8 could be used. It will be seen that a 4 × 10 girder (opposite the figures 4992 in the column marked 10) could also be used.

Allowable Uniformly Distributed Loads for Solid Wood Girders and Beams Computed for Actual Dressed Sizes of Douglas Fir, Southern Yellow Pine (Allowable Fiber Stress 1400 lbs. Per Square Inch)							
Solid Girder Size	Span in Feet						
	6	7	8	10	12	14	16
2 × 6	1318	1124	979	774	636	536	459
3 × 6	2127	1816	1581	1249	1025	863	740
4 × 6	2938	2507	2184	1726	1418	1194	1023
6 × 6	4263	3638	3168	2504	2055	1731	1483
2 × 8	1865	1865	1760	1395	1150	973	839
3 × 8	3020	3020	2824	2238	1845	1560	1343
4 × 8	4165	4165	3904	3906	2552	2160	1802
6 × 8	6330	6330	5924	4698	3873	3277	2825
8 × 8	8630	8630	8078	6406	5281	4469	3851
2 × 10	2360	2360	2360	2237	1848	1569	1356
3 × 10	3810	3810	3810	3612	2984	2531	2267
4 × 10	5265	5265	5265	4992	4125	3500	3026
6 × 10	7990	7990	7990	6860	6261	5312	4593
8 × 10	10920	10920	10920	9351	8537	7244	6264
2 × 12	2845	2845	2845	2845	2724	2315	2006
3 × 12	4590	4590	4590	4590	4394	3734	3234
4 × 12	6350	6350	6350	6350	6075	5165	4474
6 × 12	9640	9640	9640	9640	9220	7837	6791
8 × 12	13160	13160	13160	13160	12570	10685	9260
2 × 14	3595	3595	3595	3595	3595	3199	2776

Built-up Girders: Multiply above figures by .897 when 4″ girder is made of (2) 2″ pcs.
Multiply above figures by .887 when 6″ girder is made of (3) 2″ pcs.
Multiply above figures by .867 when 8″ girder is made of (4) 2″ pcs.
Multiply above figures by .856 when 10″ girder is made of (5) 2″ pcs.

Built-up girders will carry smaller loads than solid girders. Two 2″ dressed planks equal only 3″, whereas a dressed 4″ plank equals 3 1/2″.

After the sills have been placed, the girders are assembled if of the built-up type, or framed if of the solid type. Their placement and support are very important. Temporary columns are sometimes used until the building has been framed and then permanent columns are placed. This, however, should be avoided if possible, and the girder and columns placed permanently before any framing is set upon them.

TOOLS AND EQUIPMENT FOR BUILDING AND INSTALLING GIRDERS

Crosscut saw	Level	Steel framing square
Rip saw	Chalk line	Saw horses and planks
Hammer	Steel tape	Firmer chisel

How to Assemble a Built-up Girder

Build the girder on top of the side wall, using the wall as a platform upon which to rest the girder while it is being assembled.

1. Select straight lumber free from knots or other defects. Use long lengths of stock so no more than one joint will occur in the span between bearing points.

2. If $2'' \times 10'' \times 16'$ planks are used, cut one plank 8' long. Mark the crowned edge of the plank, and place this edge at the top of the girder.

3. Nail one 16' and one 8' plank together. Place two 20d nails about 6" from the end of the plank and about 2" from the top and bottom. Drive them at an angle of about 10° into the two planks. Do not drive them home so the points protrude, as this makes it difficult to place the third plank on the girder.

4. Stagger the nails 32" apart over the entire length of the girder, working from the end and keeping the top edges of the planks flush with one another as they are fastened.

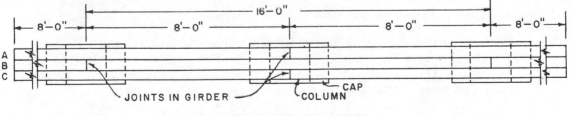

FIG. 4—16   BUILT—UP GIRDER

5. Stagger the joints in the beam as shown in Fig. 4-16. Be sure the planks are squared at each joint and butted tightly together.

6. After the first two members of the girder have been fastened, place the third member against the side of the girder opposite to that from which the nails were driven. Make sure the crown is at the top of the girder.

7. Place the nails in the same manner and proceed as when nailing the first two members together. Do not place the nails directly opposite the ones on the other side of the girder. Drive nails in solidly on both sides of the girder.

How to Frame Solid Girder Joints

NOTE: The half-lap joint is sometimes used to join solid beams. To frame the solid beam proceed as follows:

1. Place the beam on one edge, crowned side up so the annual rings run from top to bottom of beam.

2. Lay out a centerline across the end of the beam as at AB, and on both sides of the beam as at BC. Make line BC about the same length as the depth of the beam, but not less than 6″. Square lines through points C and D on both sides of the beam and connect them across the top (DD). Mark an X on one of the sections as shown. This shows the portion of the beam to be cut out.

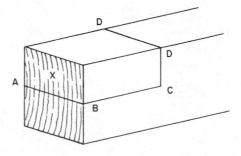

FIG. 4−17   LAYOUT OF HALF−LAP JOINT

3. Use a ripsaw and a crosscut saw to remove section marked X.

4. Test the cut surfaces with a steel square. If they are not square, pare the surfaces with a firmer chisel until they are square and true.

5. To make the matching joint on the other beam, proceed in exactly the same manner, but be sure the crowned side of the beam is down. When the joint is finished, turn the crowned side up and match the joints. Be sure all surfaces of the joint are square and tight.

6. Fasten a temporary strap across the joint to hold it tightly together. Drill a hole through the joint with a bit about 1/16″ larger than the bolt that is to be used.

7. Place the bolt and washers and tighten them.

NOTE: Provide a bearing cap at least 8″ longer than the length of the joint, Fig. 4-17. If metal is used, it should be 3/4″ thick. If wood is used, it should be not less than 2″ thick. If a butt joint is used the straps are generally about 18″ long and are bolted to each side of the beams.

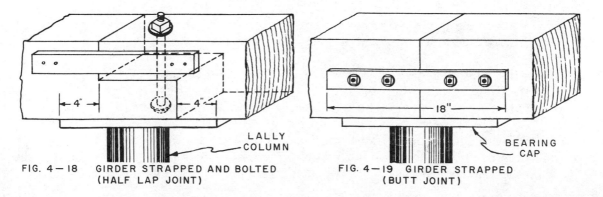

FIG. 4−18   GIRDER STRAPPED AND BOLTED   FIG. 4−19   GIRDER STRAPPED
(HALF LAP JOINT)   (BUTT JOINT)

How to Support and Erect Girders

1. Cut off the ends of the girder and frame them, if necessary, so they will fit on their permanent bearing seats in the end walls.

2. If a slight crown is desired in the center of the length of the girder, it should be considered at this time. Then cut the columns to the correct length and place them in position on the footings. Brace them in a plumb position.

3. Slide the assembled girder from the side wall, by moving both ends along the end wall to their bearing seats.

    NOTE: If the girder is long, the center will sag. This may be prevented by building staging from the side wall, on which the girder was built, to the columns. The staging should be of such a height and width that it will allow men to support the girder while it is being slid across the end walls and placed on top of the columns.

    CAUTION: Be sure to build the staging strong enough to support not only the weight of the men but also the weight of the girder.

4. Fasten the girder in place on the columns and securely brace the columns.

    NOTE: The same procedure may be used in erecting small steel girders.

REVIEW PROBLEMS   Unit 4

1. State the definition of a girder.

2. List several materials from which a girder may be made.

3. What are the requisites of a good girder?

4. The built-up girder generally used in house construction is composed of how many pieces of lumber?

5. Describe three ways in which wood joists can be framed into the side of a wood girder.

6. Why is it advisable to place a piece of 2″ stock on top of a steel girder when the joists are to rest upon it?

7. What is the minimum size ledger strip that should be fastened to the side of a girder to support wood joists?

8. Why should the ends of wood girders not be embedded in a masonry wall?

9. How much larger should a bearing plate be than the bearing of a girder?

10. What is the purpose of a bearing plate?

11. Should wood shims be used under girders to level them?

12. Where should girders be placed in house construction?

13. What is the recommended distance between girders?

14. How can the strength of a girder be increased?

15. How does doubling the width of a girder affect its strength?

16. What effect does doubling the depth of a girder have on its carrying capacity?

17. What is the most economical way to secure additional strength in a girder?

18. What proportion of the total weight of the floor does a single girder carry when the joists are lapped or butted?

19. If the joists are continuous over a girder, what proportion of the floor weight would the girder be required to carry?

20. Describe the nailing required for a built-up girder to fasten it together.

21. How much longer than the length of the joint in a solid girder should the bearing cap be?

22. If straps are used to tie a butt-jointed solid girder together, how long should they be?

23. How should the base or top of a girder be leveled?

24. What procedure would you use in erecting a girder?

25. List the tools necessary to frame and erect girders.

Unit 5   COLUMNS

It is important for the carpenter to know some of the principles that determine the proper sizes of structural members. However, it is not necessary that he understand the mathematical basis of the design of these members. In describing supporting columns, this unit will consider some of the elementary principles involved and will acquaint the learner with methods of using simple charts and data.

TYPES OF COLUMNS

A column is a vertical member designed to carry the live and dead load imposed on it. It may be made of wood, metal, or masonry. Wood columns may be solid timbers or may be made up of several wood members fastened together. Metal columns are made of heavy pipe, large steel angles, channels, or I-beams.

LALLY COLUMN   BUILT-UP WOOD COLUMN   REINFORCED CONCRETE COLUMN   I-BEAM   SOLID WOOD COLUMN

FIG. 5—1

SETTING COLUMNS

Regardless of the material of which the column is made, it must have some form of bearing plate at the top and at the bottom. This plate acts as a means of distributing the load evenly over the cross-sectional area of the column. Basement posts that support girders should be set on concrete footings. The base of a wood post should be about 3 inches above the surface of the floor to prevent the accumulation of moisture at this point. Columns should be securely fastened at the top to the girders they support, and at the bottom to the footings upon which they rest.

The solid wood column should have a metal bearing cap, drilled to provide a means of fastening it to the column and to the girder. The bottom of this type of column may be fastened to the concrete footing by a metal dowel inserted in a hole drilled in the bottom of the post and in the concrete footing or by means of a metal boot or metal strap. See Fig. 5-7, Page 39. The base at this point is sometimes coated with asphalt or pitch to prevent rust and rot.

The cap of a metal column should have holes drilled in it so that it may be lag-screwed to a wood girder. The metal cap may be bolted or welded to a metal girder. The base of a metal column should be fastened on the concrete footing and the concrete floor poured around it to keep it from moving.

LOCATING COLUMNS

It is well to avoid spans of more than 10 feet between columns that are to support girders. Many times this is controlled by local building codes. The farther apart the columns are spaced, the larger the girder must be to carry the joists over the span between the columns.

DISTRIBUTION OF LOADS ON COLUMNS

The distribution of loads on columns or posts is shown in Fig. 5-2. In this case, the girder is joined over the post B and rests on the outer posts A and C. The post B takes one-half the total girder load on each side so the load on the post is one-half the girder weight itself plus one-half the load carried by each length of girder.

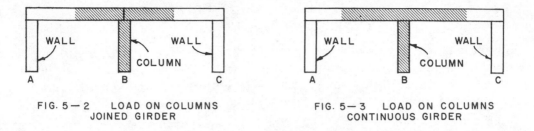

FIG. 5—2   LOAD ON COLUMNS   FIG. 5—3   LOAD ON COLUMNS
        JOINED GIRDER   CONTINUOUS GIRDER

In Fig. 5-3, the post B carries approximately five-eighths of the load on the girder from A to C because the girder is continuous above post B.

In Fig. 5-4, the load on post B is five-eighths of the total load from A to B, plus one-half of the load from B to C. The load on post C is one-half of the load from B to C, plus five-eighths of the load from C to D.

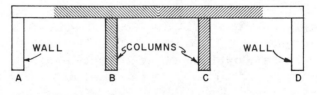

FIG. 5—4   LOAD ON COLUMNS — CONTINUOUS GIRDER

In Fig. 5-5, the load on posts B and D is similar to the loads on posts B and C of Fig. 5-4. Post C will bear one-half the weight imposed on the girder between B and C and one-half of the weight between C and D.

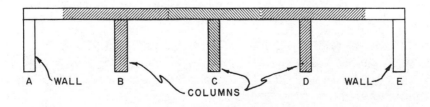

FIG. 5—5   LOAD ON COLUMNS — CONTINUOUS GIRDER

In general, where more than one post is to be used, each post must be considered separately. Although the above-mentioned rules are not exactly correct in all cases, they are practical where the posts are evenly spaced.

A METHOD OF DETERMINING THE SIZE OF COLUMNS

Assume that a two-story house is to be 24′ wide and 40′ long with an average total load on the floors of 40 lbs. per sq. ft. The spacing of the columns is to be arranged as shown.

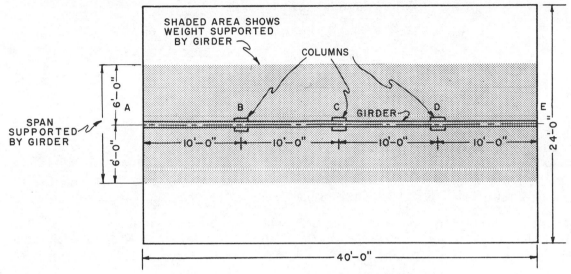

FIG. 5—6   LOCATION OF COLUMNS

1. Find the total load on the girder. (See Unit 4.)

    12 ft. (half width) $\times$ 40 ft. (length) = 480 sq. ft.

        480 sq. ft. 1st floor
        <u>480</u> sq. ft. 2nd floor

        960 sq. ft. both floors
       <u>$\times$ 40</u> lbs. per sq. ft.

    38,400 lbs. total load on girder

2. Find the load per linear foot of girder.

    38,400 lbs. $\div$ 40 ft. = 960 lbs. per lin. ft.

3. Find the load on Column B.

    This equals: 5/8 of the total load from A to B.

        10 ft. $\times$ 960 lbs. per lin. ft. $\times$ 5/8 = 6,000 lbs.
                      Plus
      1/2 of the total load from B to C

      10 ft. $\times$ 960 lbs. per lin. ft. $\times$ 1/2 = <u>4,800</u> lbs.

           Load on Column B   - - -   10,800 lbs.

4. Find the load on Column D.   Same as Column B above.

5. Find the load on Column C.

This equals: 1/2 of the total load from B to C

10 ft. × 960 lbs. per lin. ft. × 1/2 = 4,800 lbs.

Plus

1/2 of the total load from C to D

10 ft. × 960 lbs. per lin. ft. × 1/2 = 4,800 lbs.

Load on Column C   - - -   9,600 lbs.

Column B must support 10,800 lbs.

Column D must support 10,800 lbs.

Column C must support 9,600 lbs.

The loads imposed on columns or posts are called compressive and tend to bend the column unless the cross section is a certain proportion of the length. The table which follows gives the maximum loads and lengths of wood columns.

In the previous example, column B, which would have to support 10,800 lbs., would need to be a 4 × 4 according to the table, if it is to be 6'-6" high. The compressive capacities of steel beams may be found in a structural steel handbook.

Table of Maximum Loads to be Imposed on Columns Made From Douglas Fir or Southern Pine, No. 1 Grade						
Nominal Size	3" × 4"	4" × 4"	4" × 6"	6" × 6"	6" × 8"	8" × 8"
Height of Column	Imposed Load in Pounds					
4'-0"	8,720	12,920	19,850	30,250	41,250	56,250
5'-0"	7,430	12,400	19,200	30,050	41,000	56,250
6'-0"	5,630	11,600	17,950	29,500	40,260	56,250
6'-6"	4,750	10,880	16,850	29,300	39,950	56,000
7'-0"	4,130	10,040	15,550	29,000	39,600	55,650
7'-6"		9,300	14,400	28,800	39,000	55,300

TOOLS AND EQUIPMENT FOR INSTALLING COLUMNS

10'-0" pole or measuring stick   Brace and bit
Chalk line   Spirit level or transit
Hammer   Star drill
Crosscut saw   Metal dowels

It is assumed that the concrete footings upon which the basement posts are to be placed have been designed and built as described in the previous book of this series, "Concrete Form Construction". The bearing capacity of a footing should be three times greater than the load to be imposed by the column. The column, in turn, should have the capacity to safely carry the load above it.

How to Install Wooden Columns

1. Snap a chalked line across the footings to locate the center line of the columns on the footings.

2. Square the bottom ends of the wood columns or fit them to the top surfaces of the footings so that the columns will stand in a vertical position. It may be necessary to hold the post in position with braces.

3. Stretch a string from the bearing seat on the masonry walls across the edges of the temporarily placed columns.

4. Mark where the string touches the column. This is the point where the column will be squared.

5. Anchor the base of the column, using one of the techniques illustrated in Fig. 5-7.

6. Brace the columns in a vertical position.

   NOTE: Another method is to place the girder in the bearing seat, level the girder using temporary bracing, and then cut the permanent columns to a length determined by the distance between the bottom of the girder and the top of the footing.

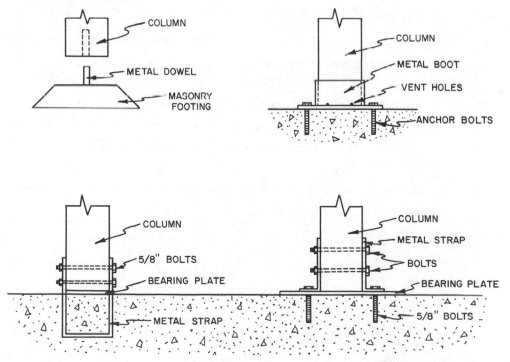

FIG. 5-7

How to Install Metal Columns

NOTE: If metal columns are used, the carpenter generally places them. The same general procedure is followed, but temporary wood columns are placed on the footings to support the girder. The precut metal columns are fastened to the girder, in their proper positions, and the bottoms are grouted in with a bed of Portland cement mortar. The finish concrete basement floor is poured around the columns to fasten them in place.

REVIEW PROBLEMS   Unit 5

1. Give the definition of a column.

2. What materials can columns be made of?

3. What is the purpose of a bearing plate at the top and bottom of a column?

4. How high above the surface of a concrete floor should a wood post be kept? Why?

5. Describe one way to fasten a wood column at the top and at the bottom.

6. Why is the bottom of a wood post, in the basement, coated with asphalt or pitch?

7. Describe one way to fasten a metal column at the top and at the bottom.

8. Why should columns be placed not more than 10 feet apart?

9. What proportion of the total load does a column under a joined girder carry?

10. Draw a sketch showing the load carried by two equally spaced columns supporting a continuous girder.

11. The compressive load on a column tends to bend it. How can this bend be avoided?

12. How much greater should the bearing capacity of a column be than the load imposed upon it?

13. Give two ways to find the length of columns when installing them.

14. List the tools necessary to install columns.

Unit 6   FLOOR JOISTS

Floor joists are perhaps the most common cause of failures in frame construction. The floor joists must be strong, stiff, and free from warp to avoid squeaking floors, sagging partitions, sticking doors, and cracked plaster. The bearing surfaces of the joists should lie squarely on the sill and girder, and they should be uniformly framed at these points so the tops of the joists form a straight and level floor line.

THE FUNCTION OF FLOOR JOISTS

Floor joists support the loads of the room they span. They usually carry only a uniform load composed of the weight of the joists themselves plus the flooring, commonly termed "dead" load, and the superimposed load such as people, furniture, etc., termed "live" load. The joists act as ties between the exterior walls and the interior partitions and tend to stiffen the frame of the building. They may also be used to support the ceiling finishes of the room below.

FRAMING FLOOR JOISTS

The bearing surface of a joist on the sill plate should be at least 4″ long and the joists should be fastened to the sill plate with at least three 8d nails, toenailed. The sill header, if used, should be fastened to the ends of the joists with at least three 16d or two 20d nails. In balloon construction, the joists should be fastened to the studs with at least two 16d nails. The joists may be framed to the girder.

Figs. 6-1 through 6-7 illustrate wood joists supported on wood girders. In Fig. 6-1 the overlapping joists are notched over the girder. The joists bear only on the ledger board, not on top of the girder, and they are toenailed into the girder with 10d nails. The joists are also fastened together above the girder with 10d nails.

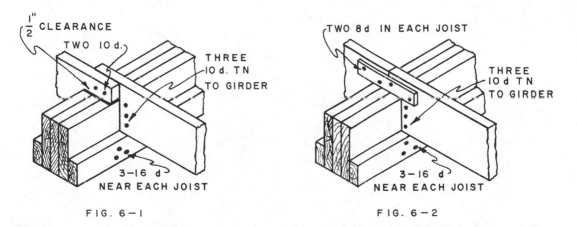

FIG. 6-1   FIG. 6-2

Fig. 6-2 shows a similar arrangement except that the joists are butted over the girder and are fastened together with an iron strap as shown. Again, the joists bear only on the ledger, not on top of the girder.

41

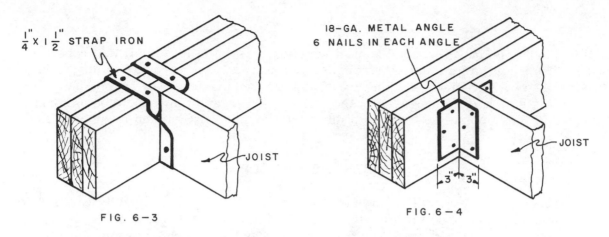

FIG. 6—3   FIG. 6—4

Fig. 6-3 illustrates the use of a bridle iron, also called joist hanger or stirrup. Both the girder and the joist are notched to accommodate the bridle iron. Fig. 6-4 shows the use of steel angles to fasten the joist to the girder.

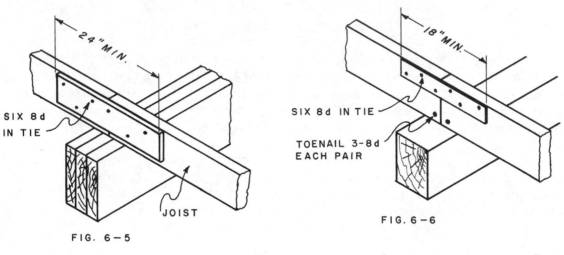

FIG. 6—5   FIG. 6—6

In Figs. 6-5, 6-6, and 6-7, the joists bear directly on the wood girder. Fig. 6-5 and 6-6 illustrate the joists butted and toe-nailed over the bearing surface, and using wood and metal ties.   Such ties may be omitted if the subfloor provides a tie across the joists.

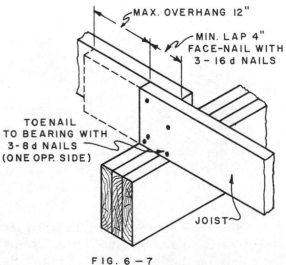

Fig. 6-7 shows the joists lapped and nailed together over the bearing surface. The minimum lap is 4" and the maximum overhang, 12".

FIG. 6 — 7

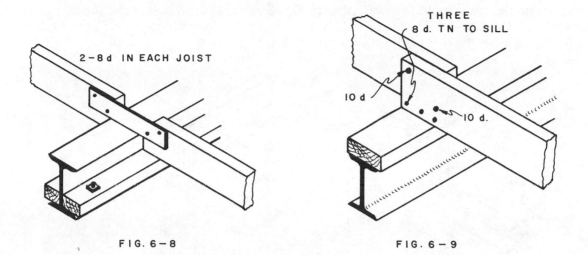

FIG. 6—8   FIG. 6—9

Wood joists which are supported on steel girders are illustrated in Figs. 6-8 through 6-16. In Fig. 6-8, the joists rest on wood ledger which is bolted, as shown, to the I-beam. The joists are tied together by means of an iron strap which is fastened across the top of the steel girder as shown. Figs. 6-9 and 6-10 show the joists lapped over the girder. In Fig. 6-9, the joists rest on a wood nailer which is bolted to beam, and are toenailed to the sill with 8d nails. In Fig. 6-10, no wood sill is used.

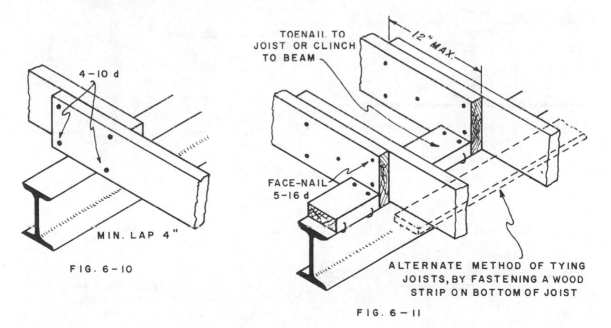

FIG. 6—10

FIG. 6—11

Fig. 6-11 shows the joists resting directly on the steel beam with wood blocking between the joists. This blocking is toenailed to the joists or clinched to the steel beam. An alternate method of tying the joists together uses a continuous wood strip fastened to the underside of the joists.

Fig. 6-12 illustrates the use of a steel angle on the I-beam upon which the joists rest, much the same as the construction in Fig. 6-8. Fig. 6-13 shows the joists resting directly on the lower flange of the I-beam. Both the latter cases use a steel strap above the girder to tie the joists together.

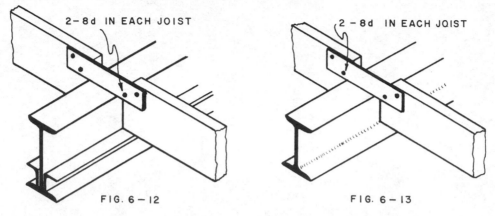

FIG. 6-12   FIG. 6-13

Figs. 6-14, 6-15, and 6-16 show cross-sections of the construction when steel beams are used.

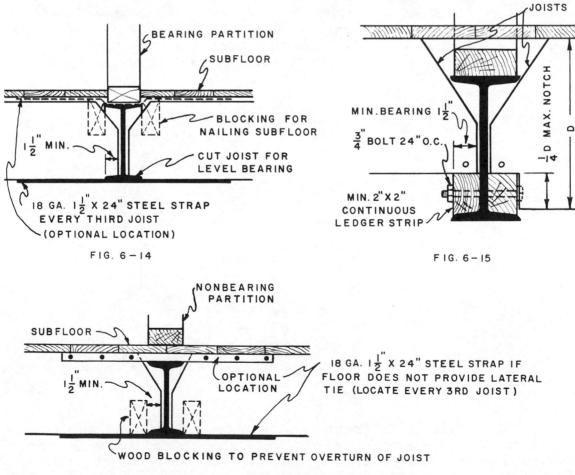

FIG. 6-14   FIG. 6-15

FIG. 6-16

COMMON CAUSES OF JOIST FAILURE

The full width of a joist should be supported at the sill and at the girder. If a joist is notched and supported as shown in Fig. 6-17, the part of the joist carrying the load, in this case, is reduced to 5 inches in width, the lower part of the joist carrying very little load. A joist framed in this manner is likely to split eventually as shown.

The bearing surface of a joist on the sill should be at least 4 inches long. If the ends of joists are lapped at the girder, they should not project more than 12″ beyond the sides of the girder.

If joists are notched where they are framed at the girder, a ledger board should be nailed on the side of the girder, Fig. 6-18. The full width of the joist (A, Fig. 6-18) will then be supported.

Fig. 6-19 illustrates an improper method of notching the joists over the ledger strip. Such a method causes the ledger strip to carry an undue load.

At the other end of notched joists, the sill should be doubled, Fig. 6-20, and the full width of the joists will then be supported by the bottom member.

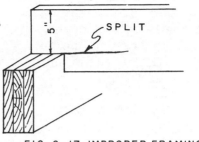

FIG. 6-17 IMPROPER FRAMING
OF A JOIST

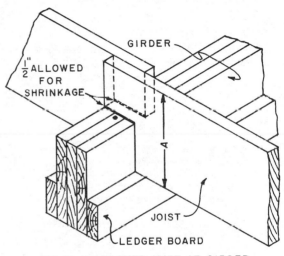

FIG. 6-18 FRAMED JOIST AT GIRDER

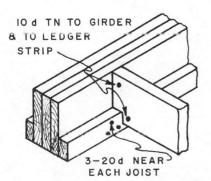

FIG. 6-19 JOIST NOTCHED
OVER LEDGER STRIP

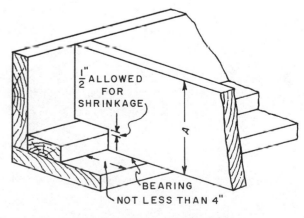

FIG. 6-20 FRAMED JOIST AT SILL

If floor joists are to be placed in a brick or concrete wall, they should have a bearing of at least 3″ and a bevel or "fire cut" on that end. In case of fire, when they burn through, the joists will fall out of the wall. Without this bevel or fire cut, the burned joist will act as a cantilever beam and in most cases cause the collapse of the masonry wall.

Floor or ceiling joists in a masonry wall should be placed in a pocket so that there is air circulating around the joist end to prevent moisture and subsequent rot of the wood member.

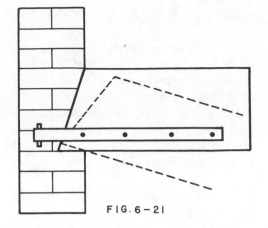

FIG. 6−21

Metal ties or anchors 24″ long are required to tie joists, which are 5 feet above grade, to the masonry walls. The anchored joists add lateral support to the wall. Wind may get behind a wall in such a manner as to force it out and cause ceiling joists to fall. The metal ties are placed near the bottom edge of the joist so, in case of fire, they will not pull the wall down, but act as a hinge and allow the joist to fall into the building. Fig. 6-22 and Fig. 6-23 show side and end anchoring of joists using metal strap and "tee" anchors, respectively.

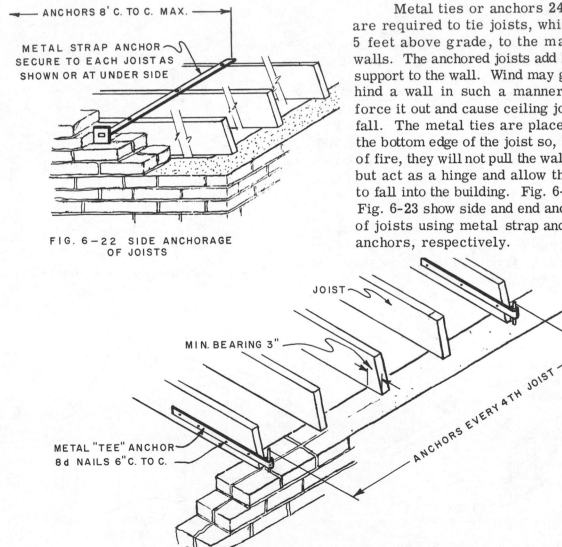

ANCHORS 8′ C. TO C. MAX.

METAL STRAP ANCHOR SECURE TO EACH JOIST AS SHOWN OR AT UNDER SIDE

FIG. 6−22 SIDE ANCHORAGE OF JOISTS

JOIST

MIN. BEARING 3″

METAL "TEE" ANCHOR 8d NAILS 6″ C. TO C.

ANCHORS EVERY 4TH JOIST

FIG. 6−23 END ANCHORAGE OF JOISTS

NOTCHING AND HOLES IN JOISTS

Notches for pipes and bored holes of the dimensions and locations as shown in Figs. 6-24 and 6-25 are permitted. In no case are the notches to be placed in both the top and bottom edges if the near sides of such notches or holes are closer than 12 inches horizontally. The notches may have a maximum depth of 1/6 of the joist depth and must be placed in the end quarter of the span, as the notches can not be placed more than 1/4 of the distance between supporting members. The holes may have a maximum diameter of 2 1/2″ and must be kept a minimum of 2″ from the top or bottom edges.

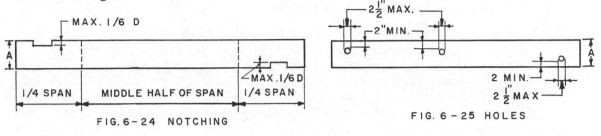

FIG. 6-24 NOTCHING   FIG. 6-25 HOLES

SELECTING WOOD JOISTS

In selecting lumber to use for floor joists, strength, freedom from warp, stiffness, and ability to hold nails firmly should be considered. When joists are being installed, the crowned edge should be placed up. If the amount of crown is excessive, the piece should not be used for a joist but may be placed in some other part of the building where it will not affect the ceiling or floor line.

The table may be used to determine the sizes of joists for ordinary houses where the load is normal. However, it is not economical to use joists over 16 feet long.

WOOD JOIST SIZES

Maximum Spans for Floor Joists - No. 1 Common

Nominal Lumber Size	Spacing on Centers	Live Load for Residential Use of 40 lb. per Square Foot Uniformly Distributed with a Plaster Ceiling					
		Maximum Clear Span Between Supports					
		So. Pine & Douglas Fir		Western Hemlock		Spruce	
		Unplast.	Plast.	Unplast.	Plast.	Unplast.	Plast.
2″ × 6″	12″	11′-4″	10′-6″	10′-10″	10′-2″	10′-4″	9′-8″
	16″	10′-4″	9′-8″	9′-10″	9′-2″	9′-4″	8′-8″
2″ × 8″	12″	15′-4″	14′-4″	14′-10″	13′-10″	14′-0″	13′-0″
	16″	14′-0″	13′-0″	13′-6″	12′-6″	12′-10″	11′-10″
2″ × 10″	12″	18′-4″	17′-4″	17′-10″	16′-10″	17′-0″	16′-2″
	16″	17′-0″	16′-2″	16′-6″	15′-8″	15′-10″	15′-0″
2″ × 12″	12″	21′-2″	20′-0″	20′-6″	19′-4″	19′-8″	18′-8″
	16″	19′-8″	18′-8″	19′-0″	18′-0″	18′-4″	17′-4″

TOOLS AND EQUIPMENT FOR INSTALLING FLOOR JOISTS

Framing square	Crosscut saw	1 1/2″ chisel
Straightedge	Steel tape or 10′-0″ pole	Ripsaw
Level	Hammer	

How to Space Joists

NOTE: When marking the spacing of joists along the sill, it will save time if a strip of wood is laid along the sill and duplicate spacing is marked on this measuring rod. This rod may be used to space the joists at the girder line and at the opposite side of the building.

1. Measure in from the outside end of the sill 3 1/2″ and then 1 1/2″ farther. This marks the inside edge of the first joist, A, Fig. 6-26. Mark an X on the left side of the mark to show where the joist is to be placed. Do likewise for the ensuing layouts.

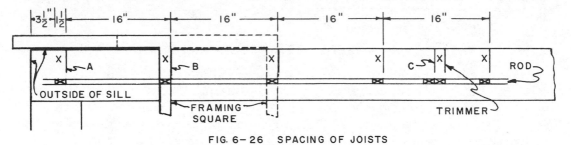

FIG. 6-26   SPACING OF JOISTS

2. Place the body of the square along the edge of the sill with the figure 16 opposite the mark, A, on the sill. Mark along the outside edge of the tongue of the square. This marks the inside edge of the second joist, B.

3. Move the square along the edge of the sill until the figure 16 on the body of the square is opposite the mark, B. Mark the line to show the position of the third joist. Continue in this manner until all joists have been located.

4. If there is a stair well or chimney well in the floor, mark the position of the trimmers wherever they come on the sill but also continue the regular spacing of the joists. This is shown at C, Fig. 6-26.

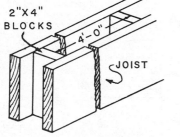

5. Determine the location of the partitions running parallel to the joists. Lay out the location of the spaced joists directly under such partitions. The two joists should be spaced apart the width of a 2 × 4.

   NOTE: This will allow the partition studs to come directly over the space and will leave room for pipes and heating ducts to pass between the joists.

FIG. 6-27   DOUBLE JOIST UNDER PARTITION

6. Mark the location of the joists on the sill on the opposite side of the building and on the girder in the same way, or use the rod if one has been made.

NOTE: If the joists are to be butted at the girder, the spacing at the girder and at both sides of the building should be started from the same point on the sill. If the joists are to be lapped at the girder, the spacing should be marked on the girder so that the joists on both sides of the building may be kept in line. Also, the spacing on the sill at the opposite side of the building should be started 1 1/2″ nearer the corner.

How to Frame Joists

NOTE: If the ends of the joists are to be notched, a templet or pattern should be carefully laid out to fit the type of sill and girder used in the building. Generally a straight joist is used for this purpose.

1. Square the ends of the joists to length.

2. Lay out the ends of the pattern joist to fit sill and girder.

3. Mark the ends of all joists from this pattern. Be sure the crowned edges will come on top.

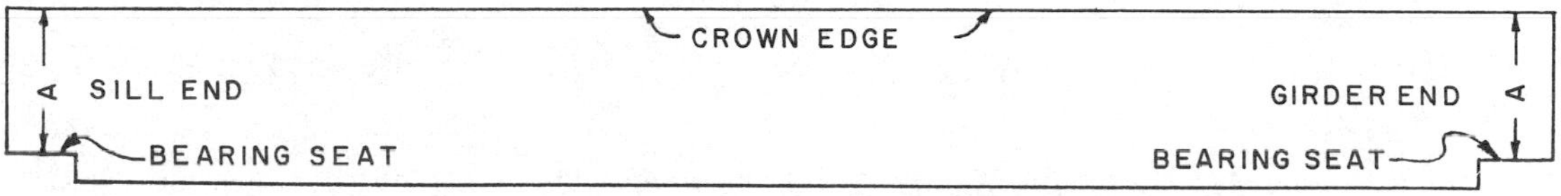

FIG. 6 — 28 FRAMING ON ENDS OF JOIST

NOTE: Fig. 6-28 shows a joist framed to fit a double sill plate and the top of the girder. In most types of framing it is important that the distances A at each end of the joist be the same; that is, the bearing seat on the girder end of the joist must be the same distance from the top of the joist as the bearing seat on the sill end.

4   Carefully saw out the bearing seat. Square the cut with a chisel if necessary.

How to Install Joists

1. Fasten two joists together, to form a spaced joist, and place it on the sill at the location of the bearing partition that runs parallel to the joists.

2. Nail all joists to the sill members and to the girder, being sure the joist rests in its proper place in line with the spacing mark.

NOTE: If the joists rest upon a sill without a sill header, they should be toenailed with at least four 8d nails, two on each side of the joist.

If the joists rest upon sills with a sill header, they should be toenailed into the sill plate. The header should be fastened to the joist with at least two 20d nails.

3. Check across the tops of the joists with a straightedge. If they are not in a straight line, find the cause and remedy it.

REVIEW PROBLEMS   Unit 6

1. State some of the results caused by inferior floor joists.

2. Why should the tops of floor joists form a straight and level line?

3. What is meant by "live load" and "dead load"?

4. Besides carrying the floor load, what other purposes do the joists serve?

5. What is the advisable area of bearing surface for floor joists?

6. How should wood floor joists be fastened to the sill plate?

7. How should the sill header be fastened to the ends of floor joists?

8. In balloon-frame construction the joists are fastened to the studs. How?

9. Sketch one method of fastening wood floor joists to the top of a wood sill.

10. Sketch one method of fastening wood floor joists into the side of a wood girder.

11. When a steel girder is used, how can the joists be fastened on the top?

12. Sketch one method of fastening wood floor joists into the side of a steel girder.

13. What is the minimum bearing a wood joist should have in a masonry wall?

14. Describe and explain the purpose of a "fire cut".

15. Do wood joists which are 2 feet above grade require metal ties?

16. How long should the metal ties for the ends of floor joists be, and how far apart should they be placed?

17. Give two reasons for using metal ties or anchors on wood joists in a masonry wall.

18. How deep and where can a notch be placed in a 2×12 joist that has a span of 12'-0"?

19. What size hole and how far in from the edge can a hole be drilled in a 2×12 joist that has a span of 12'-0"?

20. When selecting lumber for joists, what factors should be considered?

21. How can a joist with excessive crown be used?

22. Of what use is a rod indicating joist spacings?

23. Why is a mark, usually an X, placed alongside the line when laying out joists?

24. Should trimmers for stair wells or chimneys be located when laying out joists?

25. How are partitions which run parallel to the joists laid out?

26. If the joists are to be lapped over the girder, how are the opposite sides of the building plate laid out?

27. What is the last procedure when installing floor joists?

28. List the tools necessary to install joists.

Unit 7   BRIDGING

Bridging which is placed between the joists may be of three different types: wood cross bridging, metal cross bridging, and solid bridging. Bridging is used where the spans of the joists are over 9'-11". Where the span is over 9'-11", one row is used; if the span is over 19'-11", two rows are used.

THE FUNCTION OF BRIDGING

Bridging acts as a reinforcement for the joist and serves to distribute the concentrated load over several joists. The bridging adds strength to the floor by making a row of joists act as a unit in supporting the load imposed upon the floor. Bridging must be placed before the subfloor is laid, as it would be impossible to fasten the upper ends later.

TYPES OF BRIDGING

Wood cross bridging, which is commonly used, must be at least 1 × 3-inch boards nailed with two 6d nails at each end. It should be placed in double rows crossing each other from the top of one joist to the bottom of the next. The rows should be continuous from one side of the building to the other.

The practice of leaving the lower ends loose until the subfloor is laid is advisable, as this tends to bring the tops of the joists into alignment. This would not be possible if the subfloor were placed after the bridging was nailed.

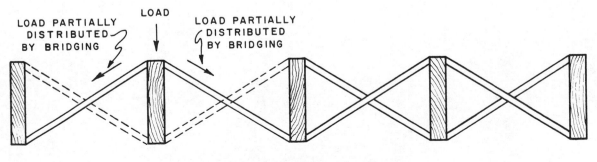

FIG. 7-1   WOOD CROSS BRIDGING

This type of bridging acts to distribute the load placed upon one joist by transferring part of it to the joists on each side by compression.

Metal cross bridging is becoming quite popular because it reduces labor costs. It is bought made up for the particular joist size and spacing. This bridging is made of at least 18-gage ribbed steel, 3/4" wide, with flat ends 2 1/2"long. Being only about 1/16" thick, it can be nailed to the top of the joist and will not interfere with the subflooring. The ribbed section adds stiffness and strength to the bridging. This type of bridging distributes the load placed upon the joist by tension to the other joists.

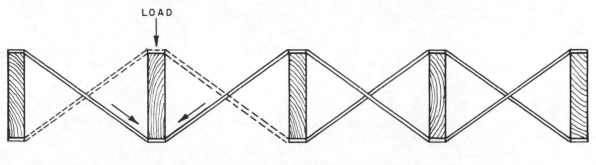

FIG. 7-2   METAL CROSS BRIDGING

Solid bridging insures a more rigid, less vibrating floor because the members are the same depth and thickness as the floor joist and are cut square and tight between the joists. Solid bridging is usually installed in offset fashion to permit toe-nailing or end nailing. This type of bridging also acts as a fire stop and is sometimes placed over the girders.

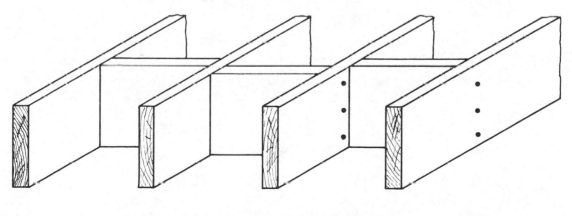

FIG. 7-3   SOLID BRIDGING

ESTIMATING BRIDGING

To estimate the amount of wood cross bridging required for a floor, measure the length of the space in which the bridging is to be placed and multiply this by three.

● Example: In a house 24′ wide × 54′ long, the girder is placed 12′ from each side wall. The span of the joists on each side is 12′, so each span would require one double row of bridging. Thus each row of bridging would be 54′ long.

$54 \times 2 = 108$ total feet to be bridged;
$108 \times 3 = 324$ lineal feet of bridging stock required.

To estimate the number of metal cross bridges required for a floor, measure the length of the floor to be bridged, then divide by 4, multiply by 3. This is the number of joist spaces if spaced 16″ o.c. Multiply this number by two to find the number of pieces of bridging required for each row. Multiply this number by the number of rows required.

● Example: Each row of bridging required is 54′ long: $54 \div 4 = 13$; $13 \times 3 = 39$; $40 \times 2 = 80$ pieces for each row: $80 \times 2 = 160$ pieces of metal bridging required for 2 rows.

To estimate the amount of solid bridging required for a floor, measure the length of the space in which the bridging is to be placed and multiply this by the number of rows required.

● Example: Each row of bridging required is 54 feet long. 54×2 equals 108 lineal feet of the same stock as joists.

TOOLS AND EQUIPMENT FOR INSTALLING BRIDGING

Framing square	Crosscut saw	1 1/2″ chisel
Straightedge	Steel tape or 10-ft. pole	Ripsaw
Level	Hammer	

How to Cut Wood Cross Bridging

1. Measure the actual distance between the joists. This distance, when the joists are 1 1/2″ thick and spaced 16″ o.c., should be 14 1/2″.

2. Measure the actual height of the joist. If 2″ × 10″ joists are being used, the actual height would be 9 1/2″.

3. Lay the square on the edge of a piece of bridging stock.

4. Hold the 9 1/2″ mark of the tongue of the square on the lower edge of the stock and the 14 1/2″ mark of the body of the square on the upper edge of the stock.

5. Mark a line across the stock along the tongue of the square and mark a point on the top edge of the stock at the body of the square.

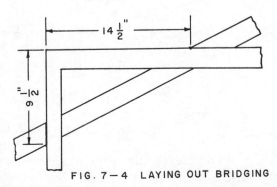

FIG. 7—4 LAYING OUT BRIDGING

6. Reverse the square, keeping the same face up, so the 14 1/2″ mark on the body is where the 9 1/2″ mark was originally, and so the 9 1/2″ mark on the tongue is where the 14 1/2″ mark was originally.

7. Mark along the outside of the tongue. The two lines on the stock will be parallel.

8. Saw across the piece at these marks.

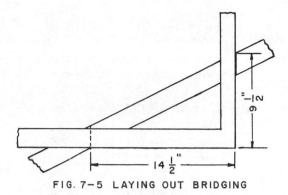

FIG. 7-5 LAYING OUT BRIDGING

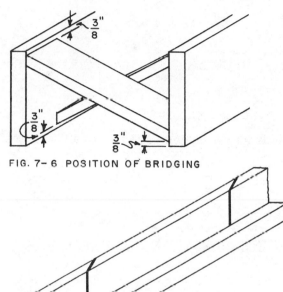

FIG. 7-6 POSITION OF BRIDGING

NOTE: Figures on the square that are about 3/8″ less than the spacing and height of the joist may be used instead of the actual sizes. This will make the bridging piece shorter so the ends will not interfere with the floor or ceiling lath.

NOTE: A miter box may be made to use in cutting the bridging.

9. To make the box, nail a piece of bridging stock to a 2 × 4 and lay out the cuts for the bridging as described in Steps 1 through 7. Cut along these lines down to the face of the 2 × 4.

FIG. 7-7   MITER BOX FOR BRIDGING

10. Place the bridging stock against the upright piece and, using the cuts as guides, saw the bridging stock to length.

NOTE: Another method of cutting bridging is as follow:

11. Cut the correct angle on one end of a piece.

12. Place this end in position against the face of a joist and at a distance from the bottom edge a little greater than the thickness of the bridging. Allow the other end to rest against the next joist and to project above the top.

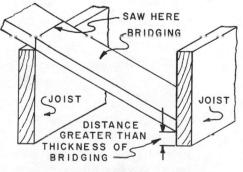

FIG. 7-8 CUTTING BRIDGING

13. Saw the bridging vertically, using the face of the joist as a guide. This method, while simple, requires somewhat more skill than the other methods.

How to Install Wood Cross Bridging

1. Start two 6d nails in ends of each piece of bridging.

2. After having determined the location of the bridging, snap a chalk line across the tops of the joists.

3. Nail one end of a piece of bridging to the side of a joist near the top and to one side of the chalk mark.

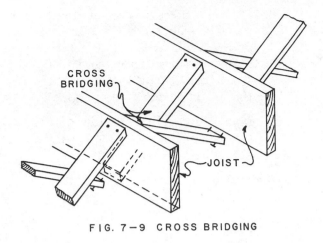

FIG. 7—9 CROSS BRIDGING

4. Nail another piece in the next opening to the same joist near the top and on the same side of the chalk mark.

5. Nail the rest of the bridging in place at the top only.

   NOTE: Do not nail the bottom of the bridging pieces until the under surfaces of the joists are ready to be enclosed.

How to Install Metal Cross Bridging

1. After having determined the location of the bridging, snap a chalk line across the tops of the joists.

2. Place bottom end down between the joists, nail the top end with two 6d nails to the top of a joist and to one side of the chalk mark. Keep the peak of the rib up to shed dust.

3. Nail another piece of bridging on the opposite side of the same joist and also on the opposite side of the chalk mark.

4. Nail the rest of the pieces of bridging in the same manner at the top only.

   NOTE: Do not nail the bottom of the bridging pieces until the under surfaces of the joists are ready to be enclosed.

How to Install Solid Bridging

1. After having determined the location of the bridging, snap a chalk line across the top of the joists.

2. One-half of the number of pieces of bridging required may be cut **14 1/2"** long if the joists are spaced **16"** on center.

3. Place a piece of solid bridging between two joists keeping the top flush with the top of the joists and to one side of the chalk mark.

4. Fasten the piece with three 16d nails through the joists.

5. Fasten the previously cut pieces in every other joist space in like manner.

6. The remaining pieces of bridging may be cut 1/16″ longer than 14 1/2″ or 14 9/16″ long if the joists are placed 16″ on centers.

7. Fasten these pieces in the remaining spaces but on the opposite side of the chalk mark.

   NOTE: These pieces can be face nailed, as they will be staggered from the 14 1/2″ pieces and will make a tight fit.

REVIEW PROBLEMS   Unit 7

1. List three types of bridging used between floor joists.

2. When is bridging required?

3. What size material is used for wood bridging?

4. Why is the lower end of bridging left loose until the subfloor is laid?

5. Explain the action of joist bridging.

6. State the advantages of metal cross bridging.

7. What gage and size material should be used for metal bridging?

8. Solid bridging can be installed in two ways. Explain.

9. What other purpose does solid bridging serve?

10. How would you estimate the quantity of wood cross bridging, metal cross bridging, and solid bridging that will be required?

11. Explain three ways of cutting wood cross bridging.

12. Why is a chalk line used when installing bridging?

13. Why is solid bridging installed in every other joist space first?

14. List the tools used to install wood bridging.

Unit 8   FRAMING FLOOR OPENINGS

When large openings are made in floors, such as for stair wells and chimney holes, one or more main joists must be cut. The location of openings in the floor span has a direct bearing on the method of framing the joists. Openings for heating and plumbing fixtures that do not require the cutting of main joists to their full depth, but only the notching of the top or bottom of the joists should be made by the carpenter, who understands joist structure, rather than by the heating or plumbing contractor.

LARGE OPENINGS IN FLOORS

The framing members around large openings in floors are generally the same depth as the joists. These members are described below.

Headers are short joists at right angles to the regular joists on two sides of a floor opening. They support the ends of the joists that have been cut off, and are usually doubled unless these cut-off joists are quite short, in which case a single header may be used.

Trimmers at floor openings are joists that support the headers. They may be either regular joists or extra joists parallel to the regular joists. If the partition studs of a stair well support the trimmer joists, reinforcement of these joists is not necessary. However, in the case of the chimney hole, the trimmer joists support the headers and should be doubled.

Tail joists are joists that run from the headers to the supporting girder or side wall.

Fig. 8-1 shows a section of a typical floor platform in which a stair well and a chimney well are located.

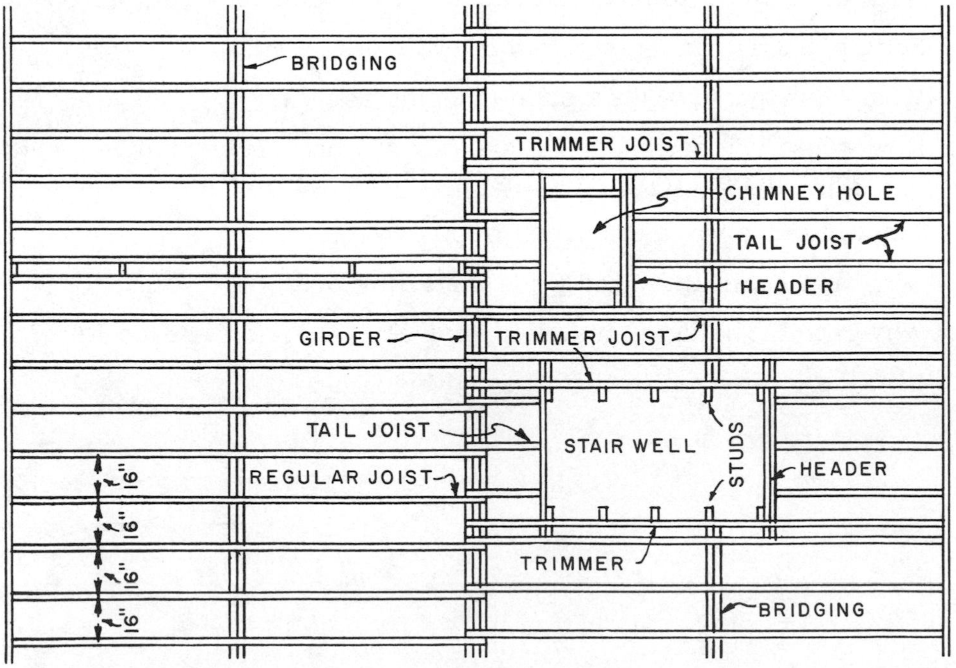

FIG. 8—1   FLOOR PLAN OF JOISTS

58

METHOD OF FRAMING OPENINGS IN FLOORS

The location of the trimmer joists should be marked on the sill and the girder at the time the other joists are spaced. The location of the trimmer joists should not interrupt the spacing of the regular joists. The location of the headers should then be marked on the trimmers on each side of the well. The inside, or single, header is then spiked between the side trimmers. The tail joists are then located according to the regular spacing of the joists and are spiked to the single header. If a double header is needed, it is spiked to the single header after the tail joists have been put in place. The small trimmers are then placed between the headers.

FRAMING A CHIMNEY HOLE

The headers, trimmers, and studs should be kept 2″ from each surface of the chimney to prevent the heat of the chimney from causing a fire. The space between the woodwork and the chimney should be filled with some incombustible material.

SIZE OF CHIMNEY WELL

The size of the chimney hole is generally figured from the flue lining. The opening in the floor joists for a chimney having one 8″ × 8″ flue would be figured as follows:

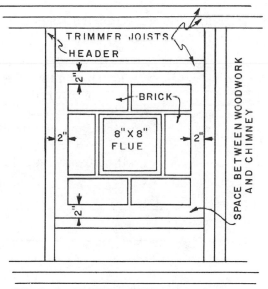

FIG. 8-2   CHIMNEY OPENING IN JOIST

The outside dimensions of an 8″ × 8″ flue are approximately 9 1/4″ × 9 1/4″. The brickwork surrounding the flue, in most cases, would consist of bricks 8″ long. Therefore, two bricks would be required to cover one side of the flue, allowing 1/2″ for the joint between the bricks. The size of the chimney opening would be 16″ for the two bricks plus 1/2″ for the mortar joint plus a 2″ space on each side of the chimney, or a total of 20 1/2″. Since the chimney is square, the hole in the floor should be 20 1/2″ × 20 1/2″. This procedure may be used for larger chimneys, but larger flues, more bricks and more joints would necessarily have to be figured to obtain the overall size of the brickwork.

FRAMING A STAIR WELL

In framing a stair well, the trimmers should be supported by the studs of the walls of the well hole. These studs should project beyond the inside face of the trimmer. Allowance should be made for this stud projection at the time of placing the trimmers. The headers and tail joists should then be placed between the trimmer joists in the same manner as for the chimney hole. See Fig. 8-1.

SIZE OF THE STAIR WELL

The length of a stair well depends on the pitch of the stairs, the type of stairs, and the headroom required. This headroom is measured from the top side of the stairs to the ceiling line above the stairs and is determined by the location of the header. An approximate method of finding the location of this header will be sufficient at this time. This method provides for a headroom of 6'-8", which is a minimum for main stairs. The tread of a main stair should be at least 9" with a nosing of 1 1/8" and a riser of not more than 8 1/4".

1. Add to the headroom the width of the floor joist.

2. Divide the sum by the height of one riser.

3. The result will be the number of risers that must be left uncovered by the construction above. This is also the number of treads that must be left uncovered.

4. Multiply this number by the width of one tread to find the length of the stair well.

● Example: Assume that the rise between the floors is 8'-4" and that the joist is 10".

   1. 6'-8" headroom desired, plus 10" for joist equals 90".
   2. 90" divided by the height of a riser of 8" equals 11 risers to be left uncovered.
   3. This is also the number of treads to be left uncovered.
   4. 11 treads uncovered multiplied by 9" for each tread equals 99" or 8'-3". Therefore, the length of the stair well between the inside faces of the headers is 8'-3".

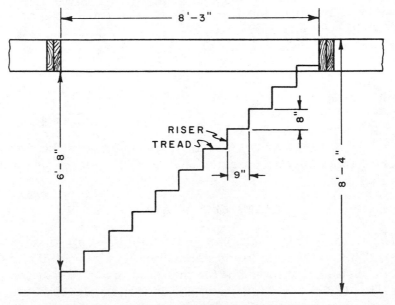

FIG. 8-3   FINDING HEADROOM FOR STAIR WELL

The width of the stair well is dependent on the width of the stairs and the allowance made for the partition studs, lath, and plaster on each side of the well.

● Example: Assume that the overall width of the stairs is 3'-2". The distance between the trimmers would then be figured as shown.

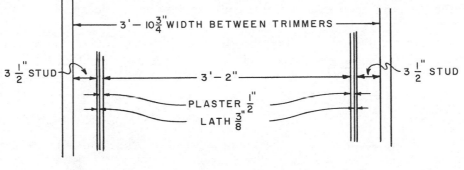

FIG. 8-4   FINDING WIDTH OF WELL HOLE

FIREPLACE HEARTHS

The wood floors in front of a fireplace should be protected by a hearth to keep them from becoming scorched. The hearth, both inner and outer, should be of noncombustible heat-resistant material. The hearth extension should be properly supported with noncombustible material. The wood forms should be removed. The outer hearth should extend at least 16" in front of the opening and at least 8" on each side. When the fireplaces have openings on more than one side, the inner hearth should be depressed, or provided with a curb at least 3" high.

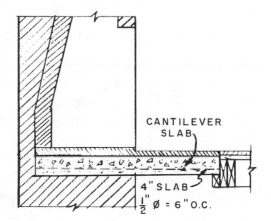

FIG. 8-5

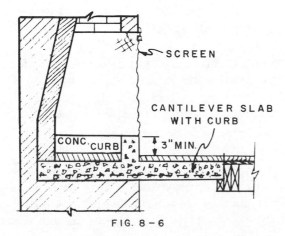

FIG. 8-6

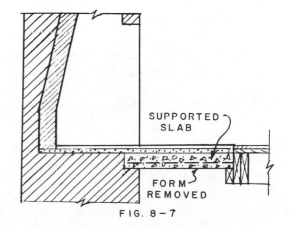

FIG. 8-7

SMALL OPENINGS IN JOISTS

If small holes, such as for pipes, are to be made through joists, they should be located halfway between the top and bottom edges of the joist. If the hole is located in the center of the joist, no material strength is lost, because the upper and lower fibers of the wood have not been cut. However, if the hole is cut through the top or bottom, the supporting fibers are cut, thus weakening the joist.

If a hole or notch of any considerable size is to be cut, the ends of the joist at the openings should be supported by a header the full depth of the joist.

TOOLS AND EQUIPMENT FOR FRAMING FLOOR OPENINGS

Steel square	Crosscut saw	Straightedge
Hammer	Steel tape	

How to Frame an Opening for a Chimney or Stair Well

A definite procedure in cutting and assembling the various joist members is necessary in framing openings. This sequence of operations is about the same in most cases. Therefore, such typical openings as those for soil pipes, chimneys, and stairs will serve as examples for all others.

1. Assuming that the opening is in the first floor, lay out the width of the well on the sill and on the girder.

2. Place the trimmers at these locations.

3. Mark the length of the well on both trimmers, allowing for double headers if they are needed.

4. Cut the headers, taking the length at the sill or girder.

5. Spike the single headers between the trimmer joists at both ends of the well. Use three 20d spikes for 2 × 8 joists and four 20d spikes for 2 × 10s.

6. Mark the regular joist spacing on the single headers and spike the tail joists to the headers at these points.

7. Spike the double headers to the single headers.

   NOTE: Short trimmers are used at the chimney hole. They are the last members to be placed and should be spiked to the single header. A block is spiked to the double header, and the trimmer is then spiked to the block.

How to Cut an Opening for a Soil Pipe

1. Measure the outside diameter of the soil pipe.

2. Lay out this diameter on the face of the joist where the pipe is to cross it.

3. Square lines from these points to the edge of the joist, allowing 1/2" clearance on each side of the pipe.

4. Saw through the joist at these marks.

   NOTE: While the joist is being cut, it should be supported by strips nailed across the top of this joist and the one on each side.

5. Nail headers to the ends of the cut joist and to the regular joists on each side.

REVIEW PROBLEMS   Unit 8

1. Why should the carpenter do the required cutting and notching of floor joists?

2. What are joist headers, and what function do they perform?

3. Why should trimmers be doubled?

4. When is it not necessary to double a trimmer joist?

5. Describe the location of tail joists.

6. Do trimmer joists affect the regular spacing of floor joists?

7. Describe how trimmer joists are installed.

8. How far from masonry should the floor framing be kept?

9. What type of material is placed between the framing and a chimney?

10. Find the size of the well for a chimney with a 10" × 12" flue.

11. Should the trimmer joists be supported by studs at stair wells?

12. What is the minimum standard allowance for the headroom of a flight of stairs?

13. State the maximum height of a stair riser and the minimum width of a tread for an easy walking stair.

14. Why are small holes allowed in the center of floor joists?

15. List the tools necessary to frame floor openings.

Unit 9   SUBFLOORING

In most frame buildings, a subfloor is laid on the joists to provide safe support of floor loads without excessive deflection and to provide adequate underlayment for the support and attachment of finish flooring materials. If a wood subfloor is laid before the rain is excluded from the building, and it usually is, provision should be made for the swelling of the subfloor. Otherwise, the exterior walls may crack or be pushed out, and other damage may be caused. To avoid this difficulty, spaces are left between the boards. Better yet, they may be protected by tarred felt paper or by sheets of polyethylene.

FUNCTION OF SUBFLOORS

An important function of the subfloor is to provide a floor during the early stages of construction as a safe platform for the workmen during the framing of the side walls. A subfloor makes the floor more substantial, especially if it is laid diagonally to the floor joists. It also acts as bracing for the joists and allows the finish wood-strip flooring to be placed either parallel or perpendicular to the floor joists. The subfloor makes the building more fire-resistant, more soundproof, and also warmer.

MATERIALS USED FOR SUBFLOORING

Wood boards with a minimum thickness of 3/4" and a maximum width of 8" may be used, with the ends cut parallel to and over the centers of the joists. This type of subfloor should be nailed to each joist bearing with 8d common nails. Two nails should be placed in 4" and 6" boards and three nails in 8" boards. Where sill headers or header joists are not used, blocking should be installed for nailing the ends of diagonal subfloor. A clearance of 1/2" between the subfloor and masonry should be provided.

Plywood of 5-ply structural grade, either interior or exterior type, 1/2" or thicker, may be used. The plywood should be installed with the outer plies at right angles to the joists and the end joints staggered. This material should be fastened to the joist at each bearing with 8d nails spaced 6" on centers along the edges and 10" on centers along the intermediate members. A clearance of 1/2" between the plywood and masonry should be provided.

Plank subfloor, when used, must be tongue-and-grooved or splined, with a nominal thickness of 2" and a maximum width of 8". The plank should be continuous over at least two spans. All the joints must be cut parallel to and over the center of the floor beams. The maximum span of 2" plank should be 7'. The planks should be blind- and face-nailed with 10d and 16d nails. Two nails should be placed in 4- and 6-inch planks and 3 nails in 8-inch planks.

ESTIMATING QUANTITIES OF SUBFLOORING

The amount of subflooring needed is estimated by adding a certain percentage to the total area to be covered. For 1″ × 6″ matched subflooring laid at right angles to the joists, add 20% to the total area. For 1″ × 8″, add 16%. Unless the floor openings are very large, they need not be deducted. If the flooring is laid diagonally, add 25% for 6″ boards and 21% for 8″ boards.

The amount of plywood needed is estimated by dividing the number of square feet in the total area by the number of square feet in the size of sheet which is going to be used. This will give the number of sheets required. A waste of about 10% is usually allowed on any type of sheet or panel.

LAYING SUBFLOORING

It is assumed that the joists, the trimmers at the well holes, and the end joists at the sill line have been aligned and fastened in their proper places. The tops of all bridging must have been nailed, and the proper reinforcing between the joists to carry bearing partitions or posts must have been installed.

Subflooring of wood boards is laid at right angles to the joists where economy of material and labor is the main consideration. It does not require the amount of cutting that a diagonal subfloor does, and consequently the labor and waste are less. With this method, it is necessary that the finish floor be laid parallel to the joists. If the finished floor were laid parallel to the subfloor, the shrinkage of the subfloor would tend to pull the finish floor boards apart.

The diagonal method of laying subflooring causes more waste and requires more labor, but it has advantages that should not be overlooked. It braces the floor and the building more rigidly than the other method, particularly when the first and second floors are laid on opposite diagonals to each other. The joints in the finish floor run opposite to the joints in the subfloor, thereby preventing the shrinkage of the subfloor from affecting the joints in the finish floor to any great extent. Diagonal subflooring also makes it possible to lay the finish floor in either direction.

TOOLS AND EQUIPMENT FOR LAYING SUBFLOORING

Hammer	Rule	Combination square
Crosscut saw	Framing square	Chalk line

How to Lay a Subfloor at Right Angles to Joists

1. Locate the outside door openings, and provide a block on which to nail the ends of the subfloor. Keep the flooring back to allow for the door sill.

2. Select several straight flooring boards with which to start.

3. Start at one side of the building and at the left end sill and cut a board so the joint will come on the center of a joist.

4. Cut another board to butt against the end of the first board and so the other end comes on a joist. Continue in the same manner until the other end of the building is reached.

5. Nail the boards temporarily so the groove side is flush with the outside of the sill.

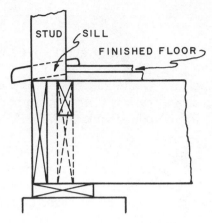

FIG. 9—I   BLOCK AT DOOR OPENING

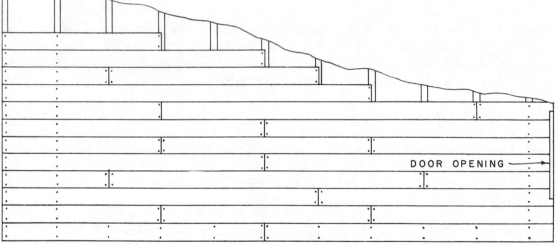

FIG. 9—2   SUBFLOOR LAID AT RIGHT ANGLES TO JOISTS

6. Align the boards from end to end.

7. Face-nail this first row of boards to each joist with 8d common nails.

8. Lay the second row of boards the length of the floor. Be sure to break the joints on different joists.

9. If the boards do not come together easily, toenail the board into the joist. Be sure to clean off the tongue of the board when it becomes bruised from the hammer.

   NOTE: Do not do any more toenailing than necessary, since this will make it difficult to draw up the next board.

10. Lay five or six rows in this manner. Then face-nail them with two 8d nails in each board at each joist.

11. Repeat these processes until the floor has been completed.

How to Lay a Subfloor Diagonally to the Joists

1. Lay out a 45° angle across the joists by measuring equal distances along the sills (AB and AC).

2. Snap a chalk line from B to C on the top of the joists.

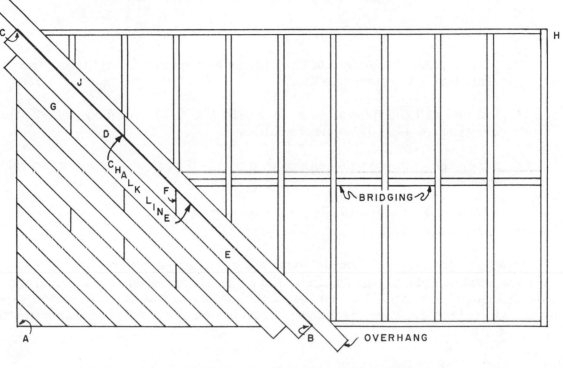

FIG. 9 — 3   SUBFLOOR LAID DIAGONALLY TO JOISTS

3. Cut a 45° angle on the end of the first piece (D). Use the combination square for laying out this angle.

4. Set this piece on the joists with the grooved edge at the chalk line and the bevel cut centered on that joist which will give the least amount of waste at the sill.

5. Nail this piece to each joist and sill with two 8d nails.

6. Cut a 45° angle on the piece E so that it can butt against the bevel cut on the first piece, as shown at F, and nail this piece in place.

7. Cut the bevel on piece G, using a shorter piece than D to break the joints. Nail this piece to the joist.

8. Fill in the rest of this strip in a similar manner.

9. Lay several more strips and then face-nail the group as previously described. Break the joints so that they will not occur in several boards on the same joist.

   NOTE: Then the overhang at each end may be cut off, or it may be all cut off after the entire floor has been laid. The overhang of subflooring at openings, such as chimney holes and stair wells, should also be cut off. A portable electric saw is a great help in these operations.

10. Continue to lay the flooring until the corner A is reached.

    NOTE: After this section of floor is laid to the corner, start to lay from the chalk line in the other direction.

11. Cut and nail the first piece in place (F, Fig. 9-3) with the tongued edge to the grooved edge of the first section.

12. Fit in the remaining pieces of this strip. Break all the joints as before mentioned.

13. Complete laying the floor to the corner H in the same manner as the first section was laid.

To lay a plank floor, follow the same directions as for the other strip flooring. The nails used in a plank floor should be 10d for blind nailing and 16d for face nailing. Plank flooring is usually laid at right angles to the beams with the underside exposed to the room below; therefore, the flooring should be selected and laid with this in mind.

How to Lay a Plywood Subfloor

1. Locate the outside door openings, and provide a block on which to nail the ends of the plywood. A section of the panel should be cut back at these locations to allow for the door sill.

2. Start at one side of the building and at the left end of the sill and cut a panel so that the joint will come on the center of a joist with the top and bottom grain of the panel at right angles to the joists.

3. Nail the panel temporarily so the long side is flush with the outside of the sill header.

4. Select another panel to butt against the end of the first panel so that the other end comes on a joist. Continue in the same manner until the other end of the building is reached.

   NOTE: The joists usually are placed 16" on centers; therefore, 8' panels could be used economically, as the ends would come over the centers of the joists.

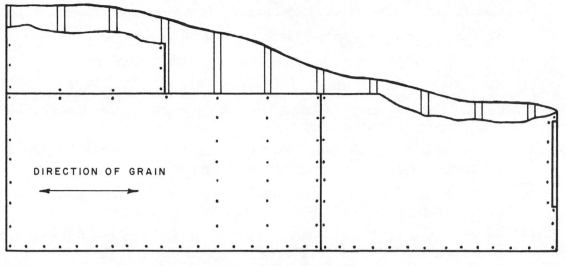

FIG. 9 — 4   PLYWOOD SUBFLOOR

5. Nail the panels temporarily so the edges are flush with the outside of the sill header.

6. Align the panels from end to end.

7. Face-nail this first row of panels to each joist and the sill header using 8d nails. Keep the nails 6" on centers along the edges and 10" on centers along the intermediate joists.

8. Lay the second row of panels the length of the floor. Be sure to break the joints on different joists and nail temporarily.

9. Then face-nail the panels to each joist.

10. Repeat these processes until the floor has been completed.

REVIEW PROBLEMS   Unit 9

1. State the purpose of a subfloor.

2. What provisions can be made for the swelling of wood subfloors before the roof covering has been installed?

3. What is the function of a subfloor?

4. State the maximum width boards that should be used for a subfloor.

5. Describe the nailing of a subfloor when 6"-wide boards are used.

6. What is the minimum thickness and what type of plywood is recommended to be used for subfloors?

7. How should plywood subfloors be applied and secured?

8. How far away from masonry should subflooring be kept?

9. What type of matching should plank subflooring have?

10. State the maximum span recommended for 2"-thick subflooring.

11. How should plank subflooring be fastened to the beams?

12. Explain the advantages and disadvantages of wood-strip subflooring when laid at right angles and diagonally to the joists.

13. What amount of waste is allowed for 1 × 6 matched subflooring, when laid at right angles and when laid diagonally to the joists?

14. What amount of waste is allowed for 1 × 8 matched subflooring, when laid at at right angles and when laid diagonally to the joists?

15. Explain how to estimate the amount of plywood that is needed for a subfloor and what amount of waste is allowed.

16. Why is it advisable to do as little toenailing as necessary when laying a strip subfloor?

17. State the advantages of starting near the middle of a platform when laying strip subflooring diagonally.

18. Why are eight-foot panels of plywood usually used in plywood subfloors?

19. List the tools necessary to lay subflooring.

Unit 10   WALL FRAMING

The various types of house framing require different procedures of layout and erection. The distinguishing feature of balloon frame is that the side wall studs extend from the sill to the roof. The second-floor joists are carried by a ribbon inserted into the studs. The attic joists are supported by a double plate nailed on the tops of the studs.

The western or platform type of frame construction is a platform consisting of joists and subfloor that is built on the sill plate. This provides a safe and convenient place on which to raise the side walls and center bearing partitions. The studs of the inside partitions and outside walls are cut to the same length and the erection of partitions and side walls is similar. The construction of the second floor is the same as the first floor. The platform, consisting of joists and subfloor, is built on the doubled top plate of the first floor studs.

Plank-and-beam construction which uses a few larger members to replace many small pieces necessary in both the balloon and platform methods of wall framing may be used in either of these two types of wall framing.

WALL FRAMING FOR BALLOON CONSTRUCTION

Studs

In balloon framing the wall studs for both the outside walls and the bearing partitions rest on the sill plate. The studs are toenailed to the sill with three 10d or four 8d nails. The studs should be continuous without splicing and may be notched one-fourth their depth for piping or conduit, or may be drilled a maximum of 1 1/4" in 2 × 4 s.

Ribbon

The second floor or ceiling joists rest on a ribbon or ledger board. This board is usually a piece of 1 × 4 or 1 × 6 square-edge stock. The studs are notched out so that the face of the ribbon sets flush with the inside edge of the studs. The ribbon is fastened with two 8d nails at each stud. The joists which rest on it are nailed to the studs with at least three 16d nails. No ribbon need be installed on the end-wall studs, as the end joists are nailed to these studs with two 16d nails to each stud.

Top Plates

The tops of the studs are tied together by the top plate, which is doubled to straighten and stiffen the wall. These plates are of 2 × 4s lapped at corners and intersecting partitions. Splices in the lower member of the top plate must be made over studs, and at least 32" away from splices in the top member. The lower member is face-nailed to each stud with two 16d nails. The top member is face nailed to the lower member with 16d nails, 16" on centers and two 16d nails at each end. When the plates are cut more than one-half their width for piping or duct work, they should be reinforced with 18-gage steel straps.

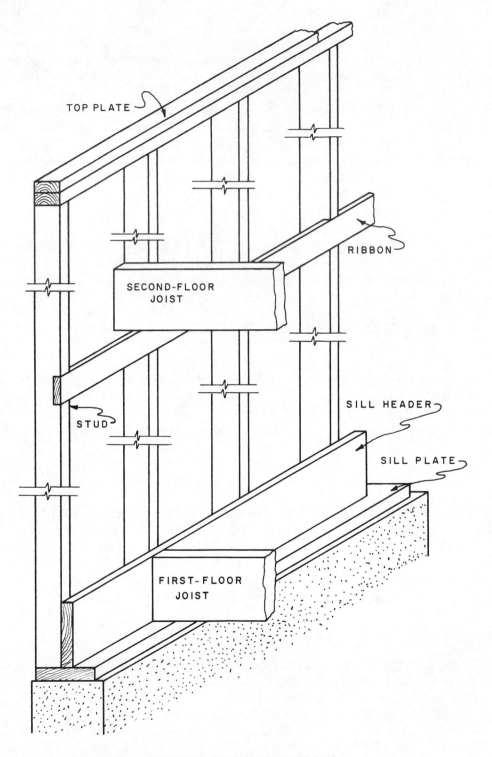

FIG. 10—1   SECTION OF BALLOON FRAME

Wall Bracing

Diagonal wall bracing is used in balloon construction unless the wood sheathing is installed at an approximate 45° angle in opposite directions from each corner or 4'-8' sheets of plywood or 25/32" fiberboard sheets are used for sheathing with nailing at both horizontal and vertical joints. Diagonal braces may be of 1 × 4 or wider boards let into either the inner or outer faces of the studs, sill plate, and top plate near each corner at approximately a 45° angle. The brace is nailed to the studs and plates with two 8d nails at each point. When an opening is at or near a corner, the full-length brace should be installed as close as possible to the opening. No knee bracing need be used.

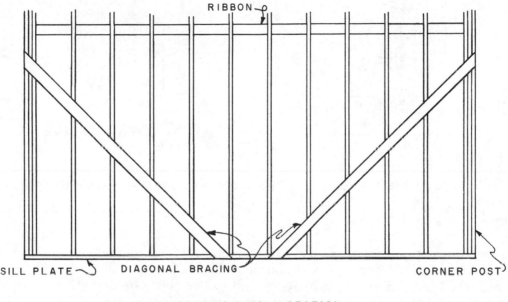

FIG. 10-2   BRACING A WALL SECTION

Corner Posts

Corner posts should be stiff and strong. When built up, they should be constructed of at least three 2 × 4s placed so as to receive the interior finish. The three 2 × 4s used should be continuous lengths. No splicing of these members is permitted, but filler blocks may be added to the posts if desired. If filler blocks are used, there should be at least three 16d nails in each block. The post members should be nailed to each other with 16d nails not more than 24" apart, staggered, on each wide face or side.

There are several different types of corner posts which are designed to give good nailing surfaces for the sheathing boards, to make the wall plumb and strong against end thrust, and to provide surfaces on which to fasten the interior wall finish.

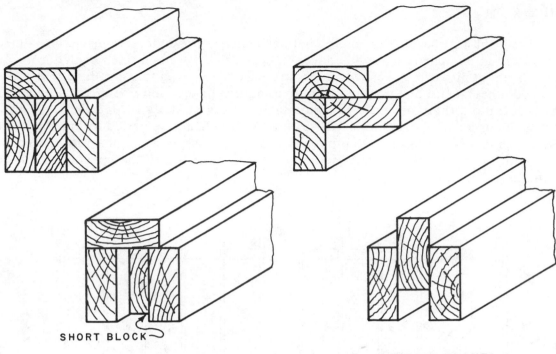

SHORT BLOCK

FIG. 10—3   CORNER POSTS FOR BALLOON AND PLATFORM FRAMES

WALL FRAMING FOR WESTERN OR PLATFORM CONSTRUCTION

Studs

In platform framing the studs of inside partitions and outside walls are cut to the same length and erection of partitions and side walls is similar. After the platform is completed the sole plate may be nailed to the subfloor and sill header with 16d nails, 16" on centers. The studs are toenailed to the sole plate with three 10d nails or four 8d nails. If the side walls are fabricated on the platform the sole plate may be nailed to the studs with two 16d nails. When the sheathing does not anchor the studs to the sill plate, additional anchorage should be provided by one-inch wide, 18-gage steel straps or special anchors designed for this purpose.

Top Plates

The top plates for platform frames are the same as for balloon frames and should be assembled in the same manner. A continuous header is sometimes used instead of top plates. The header is placed on edge; and, when it is composed of two members, it should be nailed together with two rows of 16d nails, 16" on centers. The headers are connected at the corners and intersecting partitions with sheet metal corner ties, lag screws, or other suitable fasteners. Splices in continuous headers should be staggered at least three stud spaces. The studs are toenailed to the header with three 10d or four 8d nails. The header is fastened to the corner posts with two 16d nails on each side, toenailed through the header into the post.

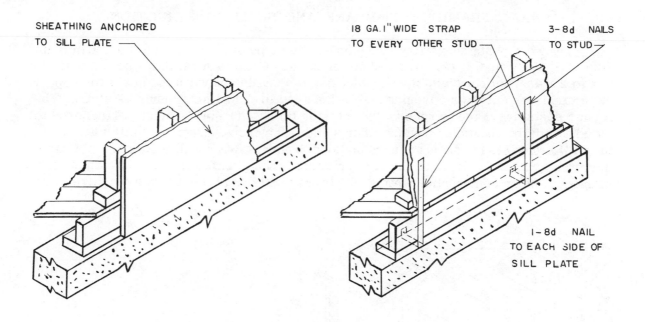

SHEATHING ANCHORED
TO SILL PLATE

18 GA. 1" WIDE STRAP
TO EVERY OTHER STUD

3-8d NAILS
TO STUD

1-8d NAIL
TO EACH SIDE OF
SILL PLATE

FIG. 10-4   ANCHORING STUDS TO SILL PLATE

Wall Bracing

Wall bracing is required in platform frames as in balloon frames. If panels are used for wall sheathing, the full-length sheets from sill plate to top plate can be installed with the grain running vertically and no additional bracing is necessary.

Corner Posts

The same types of corner posts can be used in the various types of construction as long as they are designed to give good nailing surfaces for the sheathing boards and provide surfaces on which to fasten the interior wall finish.

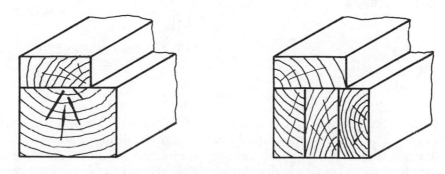

FIG 10-5   CORNER POSTS-FOR PLANK-AND-BEAM FRAME

WALL FRAMING FOR PLANK-AND-BEAM CONSTRUCTION

The wood posts supporting the beams in this type of frame should be a minimum size of 4" × 4"; and, if the posts are not more than 4' on centers, they may be built up of two 2 × 4s. Intermediate studs or blocking is installed between the posts to support the interior or exterior covering and is fastened in the same manner as studs. The corner posts are made to receive the interior finish. The posts must be anchored to the sill with the sheathing or some other suitable fasteners. When top plates are used, the wall bracing is the same as in the platform framing. When wall plates are not used, the bracing must be let into the corner posts. The greatest benefit is gained from plank-and-beam construction when 4'-0" drywall units and large glass areas are used.

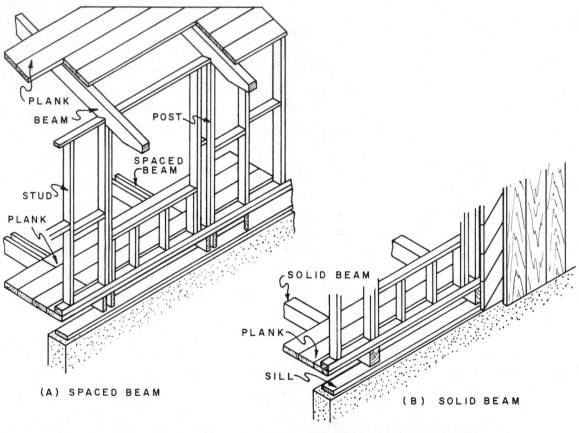

(A) SPACED BEAM

(B) SOLID BEAM

FIG. 10—6   PLANK-AND-BEAM CONSTRUCTION

Plank-and-beam construction may be combined with standard framing methods in many ways. The beams may be of solid wood timbers or may be built up of the nominal-size materials. Here are several ways that lumber can be framed to give the appearance of solid-timber construction. The spaced beam is used when plumbing or electrical work is to be concealed.

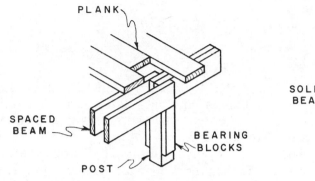

PLANK

SPACED
BEAM

POST

BEARING
BLOCKS

(A)

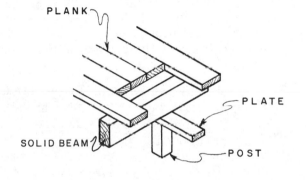

PLANK

SOLID
BEAM

BEARING
BLOCKS

POST

(B)

FIG. 10-7   BEARING OVER BASEMENT POST

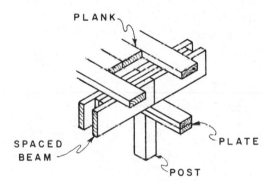

PLANK

SPACED
BEAM

PLATE

POST

(A)

PLANK

PLATE

SOLID BEAM

POST

(B)

FIG. 10-8   BEARING PARTITION 2ND FLOOR

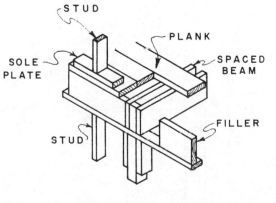

STUD

SOLE
PLATE

PLANK

SPACED
BEAM

STUD

FILLER

(A)

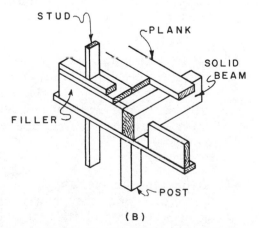

STUD

PLANK

SOLID
BEAM

FILLER

POST

(B)

FIG. 10-9 EXTERIOR WALL 2ND FLOOR

TOOLS AND EQUIPMENT FOR WALL FRAMING

Saw	Marking gage	Carpenter's level
Hammer	Steel tape	1 1/2" chisel
Straightedge	Steel square	Chalk line

How to Lay Out and Cut Side Wall Sections

The balloon frame which may be used in 1 1/2-story houses might also be called a skeleton frame, since the joists, bearing partitions, and outside wall sections are framed and erected before the sheathing or subfloor is applied. This type of construction permits rapid erection of the frame, so the roof may be placed on the building as soon as possible. Then, if the weather becomes bad, the mechanics can work inside. Sometimes the first floor is platform-framed, and then the side walls are erected with a ribbon, as in balloon framing for the second floor.

To find the length of the studs and the proper location of the notches for the ribbon, make a pattern as follows:

1. Select a stud long enough for the height of the 1 1/2 stories of the frame. (Assume that the ceiling height from the top of the subfloor to the bottom of the joists in the main floor is 7'-6" and from the top of the subfloor to the top of the plate in the second floor is 5'-6".)

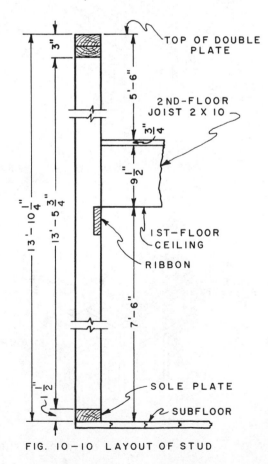

2. Starting at one end of the stud, hold the tape at 1 1/2", which is the thickness of the sole plate, and measure the height of the main story (7'-6"). This point will be the top edge of the ribbon or the bottom edge of the second-floor joists.

3. Measure up 9 1/2" for the second-floor joist and 3/4 of an inch for the subfloor.

4. From this point, measure along the stud the height of the second-story side wall (5'-6"). This point will be the top of the double plate upon which the roof rafters will rest.

5. Deduct the thickness of the double plate (3") from the point representing the top of the plate. Cut off the stud at this point.

FIG. 10-10 LAYOUT OF STUD

7'-6"	Height of first floor
9 1/2"	Width of joist
3/4"	Thickness of subfloor
5'-6"	Height of second-floor wall

12'-22 1/4" or 13'-10 1/4"   Total height of wall

3"	Thickness of double plate
1 1/2"	Thickness of sole plate
4 1/2"	Total to be deducted

13'-10 1/4"	Total height of wall
- 4 1/2"	Total to be deducted
13'- 5 3/4"	Total length of stud

6. Mark the stud for the location of the ribbon and make the notch so that the ribbon fits tightly and flush with the face of the stud.

7. Using this stud as a pattern, cut and notch the required number of studs for the side walls, including one notched stud for each corner post.

NOTE: In cutting the studs, it is a good procedure to make a templet box the length of the studs, so that all the studs may be placed in the box and cut to the same dimensions as the pattern stud.

CAUTION: Be sure to mark the bottom end of each stud, so that it will not be placed upside down in the wall.

How to Build Corner Posts

1. Select two straight studs without the ribbon notch and one with the notch.

2. Place the flat side of the notched stud on the edge of an unnotched stud. Be sure the notched edge is not covered.

3. Nail the back edge of the notched stud to the plain stud, using 16d nails spaced 24" on centers.

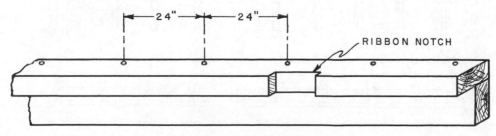

FIG. 10—11 CORNER POST

4. Place the other plain stud in the corner of these members so the side comes in contact with the notched stud and the edge touches the side of the other plain stud.

5. This allows the third stud to be nailed to the notched stud. Also the two plain studs can be fastened together, using 16d nails, 24" on center.

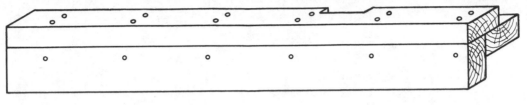

FIG. 10 – 12

6. Make a similar corner post, but with the notches for the ribbon on the opposite face of the post. These two corner posts form a pair.

7. Make another pair of corner posts in the same manner. These two pairs of corner posts are provided for the four corners of the building.

How to Lay Out the Plates and Ribbons

1. Secure a sufficient number of straight 2 × 4s to form the sole plate and single top plate.

2. Temporarily tack the sole plate to the subfloor around the perimeter of the building. The joints should be so arranged that they will come in the center of a stud, which will be over the floor joist.

3. Cut a 1 × 4 ribbon to exactly the same lengths. The joints in this member should also come on a stud.

4. Lay the single top plate and the ribbon side by side and lay out the spacing of the studs. This spacing should be the same as that of the sill plate for the joists, except at the corner of the building.

5. The location of the doors and windows should be shown on this layout.

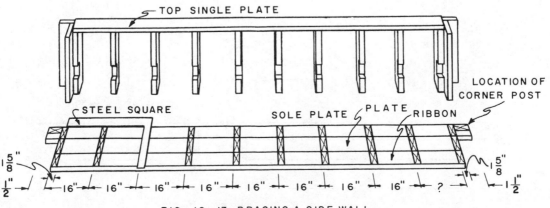

FIG. 10 – 13 BRACING A SIDE WALL

How to Assemble a Wall Section of Combination Frame

NOTE: The whole wall section or any part may be assembled and erected at one time, depending on the length of the wall and the number of helpers that can assist.

1. Lay the sole plate, with the stud location facing in, along the side of the building as near its final position as is convenient to work on the platform.

2. Lay the studs at right angles to the sill plate with the ribbon-notched edge down.

3. Lay the top plate, with the stud locations transferred to the upper edge, parallel to the sole plate.

4. Nail the sole plate to the studs and corner posts, using two 16d nails at each stud.

5. The ribbon may be tacked on the outside faces of the studs near its proper location to act as a temporary spacer if the wall section is to be covered before erection.

6. Nail the top plate to the studs in the same manner as the sole plate.

7. Any openings in the wall section may be framed at this time. This procedure saves time and labor, especially where large window openings are to be made in the walls.

8. If the wall section is to be covered before erection, the frame is squared by measuring the diagonals, and the diagonal corner braces are laid out, notched and placed before the wall section is covered.

9. If the wall sections are not covered before erection, the frame is temporarily braced. A temporary ribbon may be used to tie the studs together until erected.

10. The section is then tipped up and fastened in place by nailing 16d nails through the sole plate and subfloor into the sill header and floor joists.

11. The ribbon may be permanently fastened in place with two 8d nails at each location.

12. The corner bracing may be laid out, notched, and fastened on the outside edges of the studs, plates, and sill header.

13. If plywood, diagonal, or panel sheathing is used, the corner bracing may be omitted.

How to Assemble and Erect the End Wall Sections

1. The studs for the end walls are cut the same length as those for the side walls but are not notched to receive the ribbon.

2. Space, assemble, erect, and brace the end walls in the same manner as the side walls except for the placing of a ribbon.

   NOTE: The ribbon is not used on the end walls, as the second-floor joists will be nailed to the inside faces of these studs.

How to Cut, Assemble, and Erect the Center

Bearing Partitions

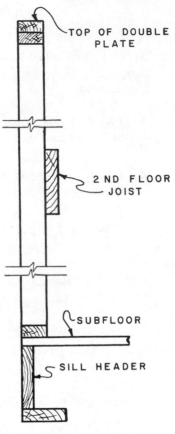

FIG. 10-14   END WALL

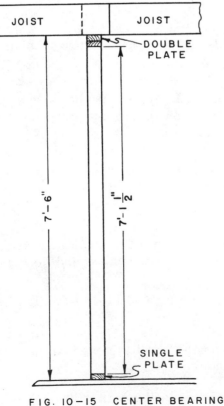
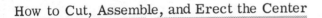

FIG. 10-15   CENTER BEARING PARTITION

1. Assume the height of the ceiling is 7'-6".

2. To find the length of the studs, deduct the thickness of the double top plate and single bottom plate from 7'-6".

3. Cut the required number of studs for the bearing partitions.

4. The center bearing partitions are assembled, erected, and temporarily braced in the same manner as the side walls.

5. The same stud length may be used for all partitions throughout the building unless the ceiling heights vary.

   NOTE: After the partition walls have been raised and braced, another 2 × 4 is nailed on the top of the plate and lapped at all corners of the wall intersections to tie the partitions together.

How to Place the Second-Floor Joists

1. Place the second-floor joists over the first-floor joists, resting them on the center partition and on the ribbon.

2. Nail the joists to the studs on the outside wall sections using at least three 16d nails at each stud.

3. Space and toenail the joists to the plate of the center partition, using three 8d nails.

4. Nail the joists at the center partition with at least three 16d nails, if they are lapped. If they are butted, nail a piece of sheathing, 24″ long, across the sides of the joists to hold them together, using six 8d nails.

   CAUTION: Do not nail the joists at the center partition unless the walls are perfectly plumb on both sides of the building.

5. The joists at the end walls are nailed to the end wall-studs with two 16d nails at each location.

   NOTE: The roof contruction is common to various types of framing methods and will be described in other units of this book.

REVIEW PROBLEMS   Unit 10

1. Is it allowable to splice wall studs? Explain.
2. How deep may studs be notched, and how large a hole may be drilled in a 2 × 4 stud?
3. What size stock is allowed for a ribbon, and how is it fastened in balloon construction?
4. Are ribbons necessary in the end walls of balloon framing?
5. What is the purpose of doubling the top plate in frame construction?
6. How should the two members of a top plate be spliced and fastened together?
7. How are stud plates, which have been cut for piping, reinforced?
8. When may diagonal wall bracing be omitted in wall framing?
9. What size of material is used for diagonal bracing, and how is it installed?
10. When no knee bracing is used, where can the diagonal wall bracing be placed?
11. How many full-length studs are required for a corner post?
12. How are corner posts fastened together?
13. Give three functions of corner posts.

14. How are sole plates fastened to the subfloor?

15. State three ways of fastening the studs to the sole plate.

16. When the sheathing does not anchor the studs to the sill plate, what other means can be used?

17. How are continuous headers fastened together?

18. Describe three ways that the corners of continuous headers may be connected.

19. How can plywood panels be applied to walls to eliminate the need for corner bracing?

20. What is the minimum size of post recommended for plank-and-beam construction when the posts are not more than 4'-0" apart?

21. When are two 2 × 4s acceptable as posts in plank-and-beam frames?

22. How are corner posts formed for plank-and-beam frames?

23. When are spaced beams used in plank-and-beam construction?

24. Why is it advisable to mark the bottom ends of studs in balloon framing?

25. Where should joints in plates and ribbons occur?

26. List the tools necessary to assemble walls in wood-frame construction.

Unit 11   FRAMING WALL OPENINGS

Substantial support must be provided around an opening in a framed wall. The type of support is dependent on the size of the opening and the load to be carried above the opening. The openings in exterior walls and in interior partitions are, in general, framed alike. The jamb studs, sometimes called jack studs, should be in one piece from the header to the sole plate. Double studs should be used at all openings to furnish the proper support to the weight above and as nailing for the trim. Although window openings under thirty-six inches wide may be single-framed, it is best to use the double framing for all openings.

The jamb stud should be nailed to the outer stud with 16d nails placed 24″ on centers, toenailed to the plate with either two 10d nails or three 8d nails, or end-nailed with two 16d nails. The outer stud should be nailed to the header with four 16d nails and toenailed to the plate with either two 10d nails or three 8d or end-nailed with two 16d nails.

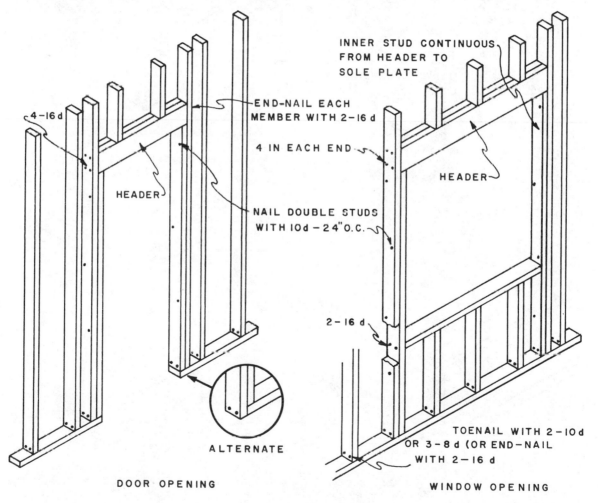

FIG. II—I   FRAMING OPENINGS IN EXTERIOR WALLS

85

The header is supported on the jamb studs and should be designed to support the load imposed upon it. As the span of the opening increases, the depth of the header increases in order to support the load. To simplify construction, the header size for the largest opening is used for all openings. This eliminates the necessity of cutting many different lengths of studs over the top of the headers.

Another method is to use a continuous header in lieu of a double top plate. Splices in a continuous header should be staggered at least three stud spaces. The members of the header should be fastened together with two rows of 16d nails placed 16″ on centers and connected at the corners and intersecting partitions with suitable fasteners, such as sheet metal corner ties or lag screws.

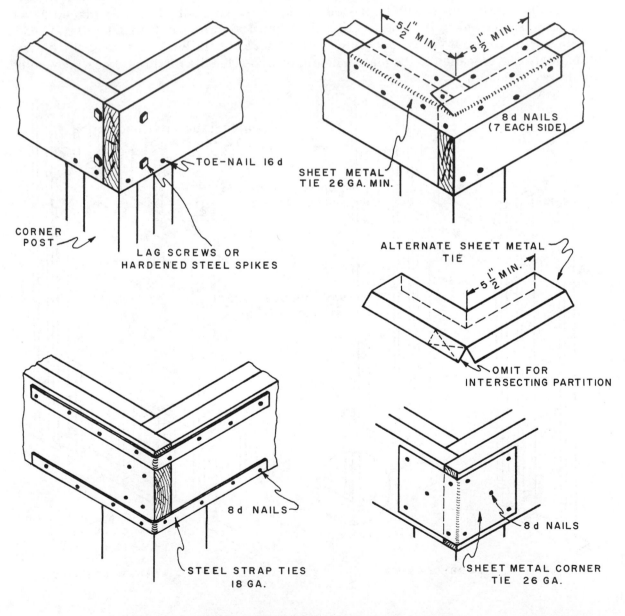

FIG. II—2 JOINING CONTINUOUS HEADERS AT CORNERS

WINDOW OPENINGS

The types of windows to be used are usually given on the elevation view or in a window schedule on the same print. This has become necessary because so many types and designs of windows are used in even the smaller homes. The window schedule, in most cases, gives the rough opening size for each type of window to be installed. If not, it will be necessary for the carpenter to refer to the manufacturer's specifications to obtain the proper size opening.

Before framing a window opening, it is necessary to know the location of the window, the width of the opening required, and the height of the opening. The location of the opening is generally given on the plans by showing the location of the center-line of the opening. The width and height of the opening are either given on the plans or can be found on the manufacturer's specifications for the window selected.

The head jambs of all windows and doors should be at the same height. The openings for windows and doors may be marked out with a story pole or window rod made from a suitable strip of wood. This rod can also be used when setting the window frames and when framing the door headers on the inside partitions. After the height of the rough opening is determined, it may be marked on the rod showing the double header over the opening and the double 2 × 4s at the bottom of the opening. If more than one size of window is to be laid out, additional double 2 × 4s may be marked on the rod. A notch should be cut out to show the exact height of the jambs at the top of the opening.

The window rod may be laid out as shown. The double header over the opening is shown at A. The height of the rough opening is shown at B, and the top jamb at C. A shallow notch should be cut here as shown. A clearance space of 1" is left between A and C. The double 2 × 4 at the bottom of the opening is shown at D. If more than one size window is to be laid out, additional double 2 × 4s may be marked as shown at E. The top of the subfloor is represented by F.

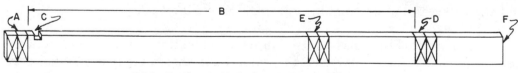

FIG II-3   LAYOUT OF WINDOW ROD

SIZE OF OPENINGS FOR DOORS

The actual size of the finished door is generally given in the plans. To find the width of the opening add to this actual size, 3/4" for each side jamb plus 1/2" on each side for the space between the jamb and stud. The height of a door opening is found by adding together the thickness of the head jamb (usually 3/4"), a 1" allowance over the head jamb, the finished height of the door, the clearance under the door and the thickness of the finish floor.

Assume that the finished door is 2'-6" × 6'-8". The width of the rough opening would be 2'-6" + 3/4" + 3/4" + 1/2" + 1/2" or 2'-8 1/2". The height of the opening would be 3/4" + 1" + 6'-8" + 5/8" + 3/8" or 6'-10 3/4". This distance would be from the top of the subfloor.

SUPPORTING THE LOAD OVER AN OPENING

There are several methods of supporting the load over window and door openings. If the opening is less than 3'-6" wide, the double 2 × 4 headers may be fastened together. This provides sufficient support for the upper joists. If the opening is more than 3'-6" wide, the size of the header should be increased to 2× 6, 2 × 8, or even more depending on the span. This is called the lintel type of support. Wider openings often require trussed headers of special design. More frequently it is possible to place a post to support the header, which, when cased, would be a mullion.

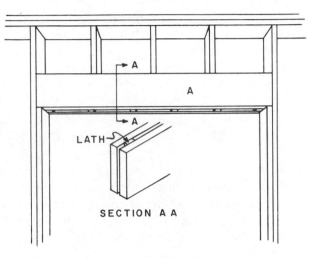

FIG. 11-4   LINTEL OVER OPENING

LAYING OUT ROUGH OPENING FOR WINDOW

Either of two methods may be used to frame window or door openings in sidewalls. In the combination frame, the studs at the openings may be left out when the wall is erected. The headers and sills of the openings are then placed in position as shown by broken lines. The short studs are next cut in and the side studs framed and placed according to the width of the opening. (See Fig. 11-5.)

Another method that may be used in the combination frame and also in the other types of frames is shown. All the studs are placed in their proper positions, regardless of the openings. The openings are then cut out and framed.

It is assumed that a combination frame is being erected and that studs have been placed every 16 inches. The opening is to be formed by cutting out parts of the studs. If the full-length studs in the openings were omitted at the time of framing the wall, the procedure is very similar except that short studs above the header and below the sill will have to be inserted later.

TOOLS AND EQUIPMENT FOR FRAMING WALL OPENINGS

Crosscut saw   Steel square   Spirit level
Hammer   Straightedge   Window rod

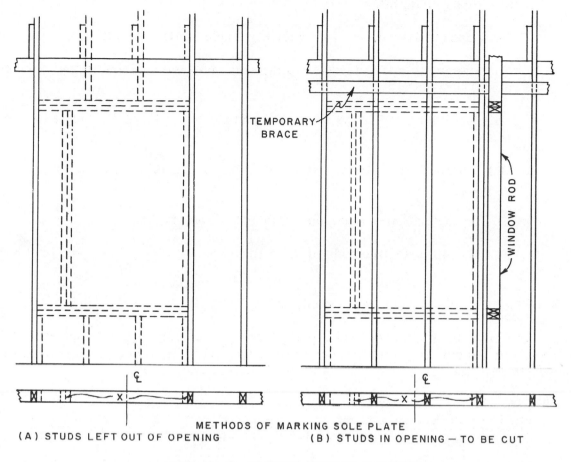

TEMPORARY BRACE

WINDOW ROD

METHODS OF MARKING SOLE PLATE

(A) STUDS LEFT OUT OF OPENING   (B) STUDS IN OPENING — TO BE CUT

FIG. 11—5   FRAMING WINDOW OPENINGS

How to Lay Out Opening for Window

1. Mark the width of the opening on the plate. Notice that **1 1/2″** is allowed on each side of the opening for the doubled 2 × 4 upright, E, Fig. 11-5 (A).

2. Place the window rod against a stud on each side of the opening and mark the location of the top and bottom headers, Fig. 11-5 (B).

How to Frame the Opening

1. Drive nails part way in at the points that show where the studs are to be cut off at the top and bottom of the opening.

2. Nail a supporting brace across four or more studs.

3. Rest a straightedge on the nails that show the cutoff mark.

4. Mark along the under side of the straightedge across the studs that are to be cut out of the wall section.

5. Saw these studs off.

6. Mark and cut four pieces of 2 × 4 for the double header and sill of the opening.

7. Nail the single sill and header to the studs on each side and to the crippled studs.

8. Nail the other 2 × 4 to the single sill and header to form the double sill and header.

9. Cut and place the trimmer studs by toenailing them to the header and sill of the opening.

10. Double the trimmer studs on the inside of the opening.

11. Nail 2 × 4s under the ends of the sill.

How to Lay Out and Frame Door Openings

The process of laying out and framing door openings is similar to that explained for windows except the bottom header is omitted and the trimmer or jack studs are placed in one piece from the top header to the sole plate.

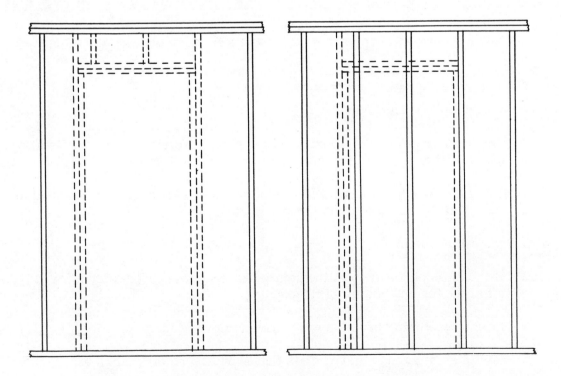

FIG. 11-6   FRAMING OF A DOOR OPENING

REVIEW PROBLEMS   Unit 11

1. What factors control the type of support needed for an opening in a framed wall?

2. Is it permissible to splice a jack or jamb stud?

3. Why are double studs used at all openings?

4. How should the jamb stud be nailed to the outer stud?

5. How is the opening header fastened in place?

6. What advantage is there in using the same size header for all openings?

7. How is a continuous header fastened together and at intersections?

8. Where does the carpenter find the rough opening sizes for windows?

9. How are the openings located on a set of plans?

10. State the advantages of using a window rod.

11. What marks are usually shown on a window rod?

12. The actual size of a finished door is given on the plans as 2'-8" × 6'-6". What should the rough opening be?

13. What is meant by a lintel type header?

14. State the maximum span recommended for a double 2 × 4 header.

15. Explain the two methods used to frame openings in side walls.

16. List the tools necessary to frame openings in walls.

Unit 12 SHEATHING

The wall sheathing, which is nailed directly to the studs, forms a base upon which to apply an exterior finish and also adds strength and insulation to the building. Several methods and materials are used for sheathing outside walls, and much technical information is supplied by manufacturers on structural qualities of their materials. However, only the fundamental qualities of sheathing suitable for frame structures will be discussed here. Wall sheathing is usually nailed to the studs, but other fasteners, such as power-driven galvanized staples, are also used.

QUALITIES OF GOOD SHEATHING MATERIAL

A. Bracing Ability

Modern methods of framing depend, to a great extent, on the sheathing for bracing. Therefore, the material used for sheathing must be strong enough to supply this bracing. The discussion in Unit 9 on horizontal and diagonal subflooring should be referred to for additional information on this topic. The holding and shearing strength of the nail which fastens the sheathing to the studs must also be considered.

B. Ability to Hold Nails

Sheathing material may well be termed subcovering for outside walls, because it is covered by clapboards, shingles, or other surface coverings. Some types of exterior coverings must be fastened to the sheathing between the studs. In these cases, the sheathing should be able to receive and firmly hold the nails used to fasten the outer wall covering. Otherwise, furring strips or other methods must be used.

C. Insulating Properties

Sheathing cuts down the heat loss through the walls of a building, thus reducing the cost of fuel in winter and keeping the building cooler in summer. The various kinds of sheathing materials vary considerably in their insulating qualities. Measures of insulating properties of various materials as worked out by manufacturers, trade associations, and government agencies may be referred to. These figures should be based on a complete wall section including both the inside and outside coverings.

These three factors, in addition to cost and ease of application, must be considered in selecting sheathing materials. One building might require exceptional bracing from the sheathing, another might demand sheathing that will hold nails well, while a third might require a material with particularly good insulating properties.

WOOD BOARD SHEATHING

When wood boards are used for wall sheathing, they should be at least 3/4" thick and not over 12" in width. If the boards are applied diagonally, extending at approximately 45° angles in opposite directions from each corner, no additional corner bracing is required. The boards should be cut with the ends parallel to and over the centers of the studs. Two 8d common or 7d threaded nails should be used in 4", 6", or 8" boards and three nails in 10" and 12" boards.

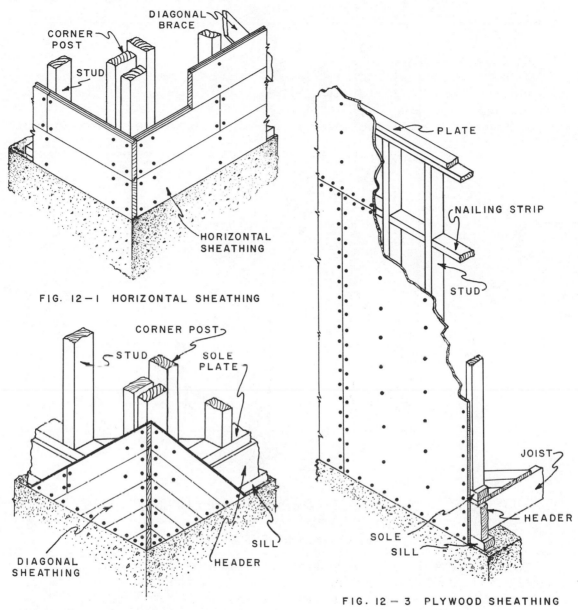

FIG. 12—1 HORIZONTAL SHEATHING

FIG. 12—2 DIAGONAL SHEATHING

FIG. 12—3 PLYWOOD SHEATHING

PLYWOOD SHEATHING

Plywood sheathing may be either of the interior or exterior type of structural grade. It should be at least 5/16" thick for studs spaced 16" on centers and 3/8" thick for studs spaced 24" on centers. Usually at least 1/2"-thick plywood is used, as the exterior finish material can be fastened to it, and the nails will have more holding power. The sheets should be applied vertically, if possible, and fastened along the edges with 6d nails spaced 6" on centers and 12" on centers along the intermediate members. No corner bracing is needed when plywood panels are used for wall sheathing.

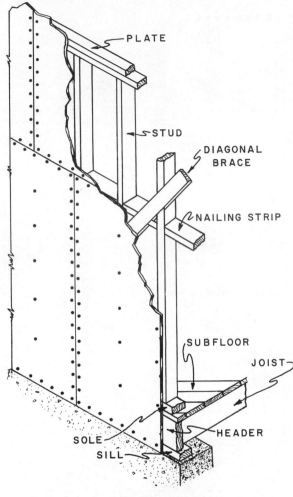

FIG. 12 — 4   FIBERBOARD SHEATHING

FIBERBOARD SHEATHING

When fiberboard 1/2″ thick is used, it is necessary to have corner bracing, but if 25/32″ thick panels are used, no corner bracing is required when the panels are installed vertically. Fiberboard sheathing cannot be used as a nailing base for exterior covering. Special methods are required to attach the exterior finish material. The 1/2″ sheets should be fastened with 1 1/2″ roofing nails and the 25/32″ sheets with 1 3/4″ roofing nails. The nails should be placed 3″ on centers around the edges and 6″ on centers at the intermediate supports.

GYPSUM SHEATHING

If gypsum sheathing is used, it is necessary to install corner bracing. The thickness of the sheathing should be at least 1/2″. Gypsum sheathing cannot be used as a nailing base for exterior covering without using nailing strips or other special methods which have been developed for attachment to nonwood sheathing. Gypsum panels should be attached to studs with 1 3/4″ nails having a 3/8″ head, spaced 4″ on centers along the edges and 8″ on centers at the intermediate supports.

TOOLS AND EQUIPMENT FOR APPLYING SHEATHING

Hammer   Level
Crosscut saw (electric or hand)   Hand axe
Chalk line   Chisel
Square   Pinch bar

How to Apply Wood-Board Sheathing

NOTE: Refer to Unit 9 for details of applying sheathing, since the operations are so similar to those of applying subflooring.

1. To apply sheathing horizontally, line up the first course of sheathing with the bottom of the sill and nail it solidly with 8d nails. All joints should be made on studs.

2. Apply the next course. Cut the boards to such a length that the joints will not come over those of the first course.

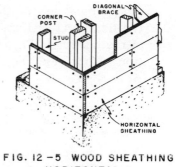

FIG. 12-5 WOOD SHEATHING HORIZONTAL

NOTE: If the waste piece cut off the end of the first course is more than 16" long, use it to start the second course, as long as the joints do not come over those of the first course.

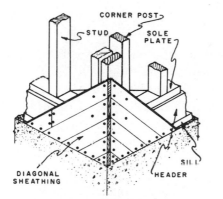

FIG. 12-6 WOOD SHEATHING DIAGONAL

3. Continue to apply the sheathing up to the bottom of the first window opening or to 6ft. above the ground.

4. At this point, consider the type of scaffold to be used, if necessary, cutting the scaffold bracket holes in the sheathing every 7 ft. Cut these holes next to a stud, not halfway between two studs.

5. Before applying sheathing diagonally, the type of scaffold to be used should be considered so as to save time later.

NOTE: There are many kinds of scaffold brackets on the market that are easily attached to the studs to support the scaffold planks. They should be well secured before workmen are allowed on the scaffold. The staging should be at least 12" wide and should be made of strong planks. Large amounts of lumber should not be piled on the scaffold but should be placed upright on the ground. Scaffolds should be placed along the side of the building no more than 7 feet above each other.

How to Sheath Around Window and Door Openings

1. Cut off the sheathing flush with the inside of the rough window or door jamb studs.

   NOTE: Some carpenters prefer to let the sheathing project beyond the rough jambs about 1″ rather than to cut it flush with the jamb. This provides for better nailing of the window frames.

2. Cut and double face-nail the sheathing along the sides of the rough jambs to the head and sill at the top and bottom of the opening. Then chop out this portion with a hand axe.

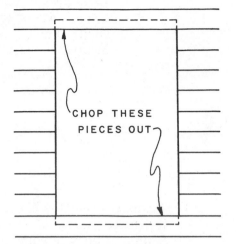

FIG. 12−7 SHEATHING AT TOP AND BOTTOM OF OPENING

How to Apply Composition Sheathing and Plywood

1. Start the first panel flush with the bottom of the sill and at the end of the wall. The other end must come on the center of a stud. Use flat-headed nails of the size and type specified by the manufacturer of the material being used.

2. Continue with more panels until the end of the building is reached.

3. Erect the panels of the second row in the same manner.

   NOTE: The piece that is cut off and left over from the first row of panels may be used to start the second row of panels, providing it is long enough to reach the center of a stud. If not, cut a full panel so that the first joint does not come directly over the first joint of the first row of panels.

REVIEW PROBLEMS   Unit 12

1. What is the purpose of wall sheathing?

2. Give the desirable qualities one should look for in sheathing material.

3. What is the minimum thickness and the maximum width recommended for wood-board sheathing?

4. Is it necessary to install corner bracing when wood board-sheathing is laid diagonally?

5. How should wood board-sheathing be applied?

6. What types, grades, and thicknesses of plywood are suggested for sheathing?

7. How should plywood sheathing be nailed to walls?

8. Can fiberboard sheathing be used as a nailing base for exterior wall covering?

9. What size nails should be used to fasten 25/32″ thick fiberboard to walls and where should they be placed?

10. How can an exterior covering be applied over fiberboard or gypsum sheathing?

11. Should wood sheathing boards be joined on a stud or halfway between studs?

12. Why do some carpenters prefer to let sheathing project beyond rough opening jambs about one inch?

13. List the tools necessary to apply sheathing to a wall.

Unit 13 SCAFFOLDS

It is difficult and unsafe for a carpenter to work on the walls of a building unless he can reach the work comfortably. Therefore, some arrangement must be made whereby he can work under safe conditions as the building increases in height. This arrangement is called staging or scaffolding. The height between lifts of a staging will vary with different kinds of work. Most carpenters can work comfortably up to a reach of six feet. The staging must be carefully planned to avoid accidents.

The designing and construction of scaffolds should be done by experienced men, and no attempt should be made to economize by the use of inferior lumber or inadequate nailing. It is important that the lumber be selected for straight grain and freedom from shakes or knots. The selection of staging material is as important as the method of erecting it. Material that is rather brittle, such as hemlock or white pine, should not be used. Spruce or fir, which are strong and tough, are very satisfactory. Green lumber should not be used because of inferior strength and hidden defects. The use of double-headed nails is a safety precaution, since they can be driven home and later may be withdrawn easily. This easy removal encourages the workmen to pull the nails out of the scaffolding when it is taken down. If the lumber is then placed on the ground, there will be no nails protruding for them to step on.

A scaffold may be defined as a raised, and usually temporary, platform for supporting workmen, tools, and some materials. Where there are more than two platform levels, requiring more substantial construction, with each upright consisting of more than one length, the structure becomes a staging as distinguished from a scaffold.

STAGING BRACKETS

The scaffolding or staging bracket is simple to make and is an excellent means of supporting a staging on a frame building. This type of bracket consists of two 2 × 4s about 4' long, two long braces of 1 × 6 about 5'-8" long, and two short gussets at the corner. A hooked bolt fastens around a stud and passes through the bracket. A nut is placed on the outer end of the bolt over a large washer to hold the assembly together.

There are several types of steel scaffold brackets on the market. These are usually slightly smaller and lighter than the wooden bracket. Some of them are made so they can be folded into a small space for convenience in storing.

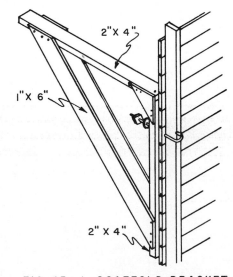

FIG. 13—1 SCAFFOLD BRACKET

THE SINGLE-POST SCAFFOLD

The single-post scaffold has an outside post only. These uprights are usually spaced from 6' to 8' apart. The uprights are usually 2×4 or 4×4 and run the entire height of the scaffold. Spruce ledger boards or outriggers are nailed to the uprights and a 2×4 cleat or block nailed on the sheathing and through it into the studding. Diagonal or X braces hold the scaffold rigid.

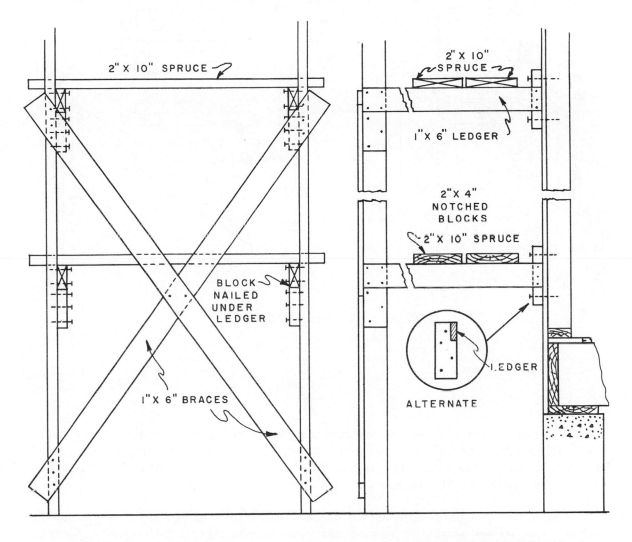

FRONT VIEW OF SCAFFOLD   END VIEW OF SCAFFOLD

FIG. 13-2   SINGLE-POST SCAFFOLD

THE CHIMNEY SCAFFOLD

The chimney scaffold is designed to straddle the roof ridge but may be made to fit one side by extending braces up to fasten over the ridge. The end sections may be made on the ground and joined on the roof. Any type of scaffold used should have a minimum safety factor ratio of four to one; that is, it should be so constructed that it will carry four times the load for which it is intended.

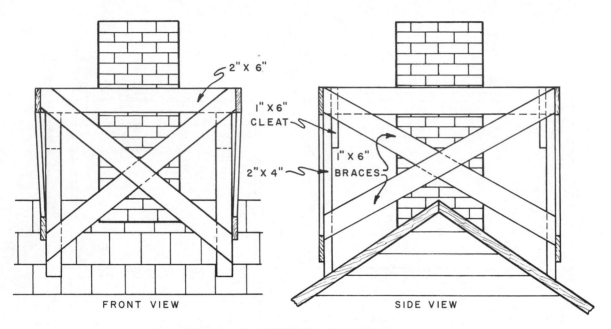

FIG. 13 — 3   CHIMNEY SCAFFOLD

PORTABLE SCAFFOLDS

Portable scaffolds can be used for many purposes in the construction of buildings. Horses may be used by workmen doing light work. Details for the construction of scaffold horses may be found in the "Safety Code for Building Construction" (American Standards Association).

Extension pieces should not be nailed on the legs of horses to increase the height. The horses should be placed on a level and solid foundation, not on bricks, tiles or other loose objects. Horse scaffolds should be made not more than three tiers or 12 feet in height. The planking should be not less than $2'' \times 10''$ and be supported by horses placed not more than 7 feet apart for light-duty scaffolds and 5 feet apart for heavy-duty scaffolds. The end horses should be placed not less than 6 inches nor more than 18 inches from the ends of the platform planks.

Overhead protection should be provided when work is carried on directly over any type of scaffold in use.

DOUBLE-POST SCAFFOLD

Some types of work require the scaffolding supports to be free from the building. An example of this is in brick veneer work. This makes a double-post scaffold necessary. This type is similar to the single-post scaffold with the addition of an inside post of the same size as the outside post. The ledger boards are nailed to the posts and should have cleats nailed under them to help carry the load. The inside post should be braced diagonally the same as the outside posts. The distance between the inside and outside posts should be at least 4 feet.

BRACING SCAFFOLDING

The braces, as well as all the other lumber used for scaffolding, should be free from large or loose knots or other imperfections that might cause them to break under strain. All the braces should be properly nailed. If the ground is soft, blocks that are larger in cross section than the posts should be placed under them to keep them from sinking in the ground.

TOOLS AND EQUIPMENT FOR ERECTING SCAFFOLDS

Hammer	Framing square	Wrench
Crosscut saw	Brace and bit	6-ft. rule

How to Erect and Secure Scaffold Brackets on Wood Sheathing

1. Determine the height of the scaffold and mark this distance on a wall stud.

2. From this point, measure down the distance between the top of the bracket and the center of the bolt.

3. At this point, bore a hole through the sheathing and alongside the stud, 1/16″ larger than the bolt.

4. Push the bolt through the hole from the inside so the hooked end will be around the stud.

5. Drive an 8d nail part way into the 2″ edge of the stud and alongside the hook of the bolt. Bend the nail over and around the bolt to hold it in place.

6. Place the bracket over the bolt on the outside of the building.

7. Put the nut and washer on the bolt and tighten the nut. After the brackets have been put in position on one side of the building, place the planks on the brackets and fasten them in place.

   NOTE: When composition sheathing is used, it is necessary to put two boards between the bracket and the sheathing. These boards should run horizontally and should cover at least two studs. They must be nailed to keep them from falling. The top piece should have a hole bored through it. to receive the bolt. These pieces will then carry the weight of the bracket, which would otherwise be forced down through the soft sheathing material.

How to Erect a Single-Post Scaffold

Assume that the building is 40′ long and 20′ high from the ground to the cornice. The scaffold will require seven uprights spaced approximately seven feet apart. Use 18′-long 2 × 4s for the uprights.

1. Measure up 6′ on the 2 × 4 upright.

2. At this point, nail on a 1″ × 6″ × 4′ spruce ledger board with four 10d nails. Make sure that the ledger board is at right angles to the upright.

3. Nail a piece about 12″ long on the upright directly under the ledger board.

4. Measure 6′ from the ground up along a stud on the building.

5. Using this as a center point, nail a piece of 2″ × 4″ × 18″ stock on the building with three 20d nails. The nails should go through the sheathing into the stud.

6. Stand the upright in place and approximately plumb.

7. Nail the ledger board to the side of the 2 × 4 on the building.

8. Prepare and erect the other four uprights in the same manner.

9. Nail on the diagonal braces.

10. Put the planks in place.

11. After the building has been sheathed up another 6 feet, erect and brace the next lift or height of staging in the same manner.

How to Erect Double-Post Scaffolding

Assume that the building is 40′ long and 20′ high from the ground to the cornice. Seven sets of uprights will be required. Each set will consist of two 2 × 4s, each 18′ long.

1. Place one 18′ upright not less than 6″ from the building and the other one 4′ outside the first one.

2. Measure 6′ up from the bottom on each post.

3. Nail the ledger board on at this point with four 10d nails in each post.

4. Nail a 4′ horizontal brace across the top of the two uprights. This is to keep the upright posts parallel.

5. Nail a block on each post directly under the ledger board.

6. Make another set of uprights in the same manner.

7. Stand the two sets of uprights in place.

8. Brace this scaffold in the same manner as the single-post scaffold with the addition of another set of the same type of braces for the inside posts.

9. Build the rest of the uprights in the same manner.

10. Place the planks in the same way as for single-post scaffolding.

11. Build each succeeding lift or height of scaffolding in a similar manner.

12. Brace the end uprights to the ends of the building.

REVIEW PROBLEMS   Unit 13

1. How far apart, vertically, should the lifts of a staging be placed?
2. What qualities should lumber used in a scaffold have?
3. State the advantages of using double-headed nails when constructing scaffolds.
4. What is the difference between a staging and a scaffold?
5. Describe a simple wooden staging bracket.
6. How far apart, along the side of a building, should the posts of a single-post scaffold be located?
7. Describe the construction of a single-post scaffold.
8. What is the safety factor ratio recommended for scaffold construction?
9. What is meant by the safety factor?
10. List three operations for which the portable scaffold may be used.
11. Explain the difference between the single-post and the double-post scaffold.
12. What precaution should be taken if the ground is soft when erecting a scaffold?
13. How can staging brackets be applied to a composition board sheathed wall?
14. Where are ledger boards located in a scaffold?
15. List the tools necessary to erect a single-post scaffold.

Unit 14   TYPES OF ROOF FRAMING

Although there are many types and combinations of types of roofs used in frame construction, the two basic types are flat and pitched. The flat roof uses larger ceiling joists as rafters. The so-called flat roof may have a slope of up to three inches in vertical rise to every twelve inches of horizontal run. The flat roof must be well supported at the walls, and the slope is usually a minimum for drainage. This type of roof must have a built-up covering of bituminous materials.

The pitched roof has a greater slope than the flat roof and is usually considered as varying from three inches to twelve inches in vertical rise to every twelve inches of horizontal run. This type of roof may have many different kinds of roof coverings except the built-up type, as the bituminous material would melt and run down the sloped surface. An understanding of the layout of a single-slope roof is basic to an understanding of the layout of other types of roofs.

TYPES OF ROOFS

The simplest roof construction is the flat roof. This type of roof usually has the roof joists laid level, with the sheathing and roof covering on the top and the underside supporting the ceiling material. Sometimes they are sloped from one-half inch to one inch vertical to every twelve inches horizontal to insure proper drainage. The roof rafters are supported by the plate on each side wall of the building and by intermediate girders if the length of span requires them.

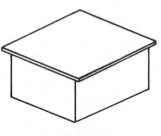

FLAT ROOF

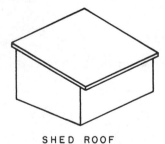

SHED ROOF

The single-pitch or shed roof is the simplest type of pitched roof, as it sheds water in one direction. This type sometimes has the ceiling finish attached to the underside of the roof rafters, and, therefore, the ceiling also has a slope to it.

The butterfly roof sheds water toward the middle of the building, which is lower than the outer edges. This style roof usually has a vertical rise of only three to four inches for every twelve inches of horizontal run. Special consideration should be given to the roof covering and flashing used on this style of roof to insure watertight construction.

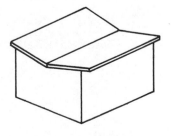

BUTTERFLY ROOF

The gable roof sheds water away from the middle of the building in two directions. The rafters are laid out in a manner similar to those of the single pitch roof, except that they must be cut for both sides of the building.

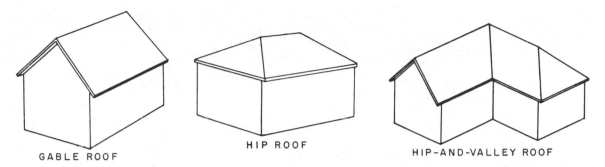

GABLE ROOF   HIP ROOF   HIP-AND-VALLEY ROOF

The hip roof slopes in four directions from each wall plate to the ridge. The same slopes are used on hip roofs as on gable roofs. The hip roof is simply two gable roofs run together at right angles to form the peak or ridge. If the surfaces terminate in a point, it is referred to as a pyramid roof.

The hip-and-valley roof, which is a combination of a hip roof and an intersecting gable roof, has many variations. Usually the intersection is at right angles; but the slopes or pitches of the roofs may be different, or the ridges may not be at the same elevations.

The gambrel roof is similar to the gable roof but has two different slopes on each of the long sides. Usually the lower slope is quite steep, and the upper slope relatively slight. This type allows more efficient use of the space under the roof.

GAMBREL ROOF

MANSARD ROOF

The mansard roof is similar to the gambrel roof except that each of the four sides has a double slope. This type of roof was introduced by the architect whose name it bears. When the upper part of this roof is flat, the roof is called a deck roof.

An equal-pitch roof is one in which the slope of the main roof is the same as the slope of an intersecting roof.

An unequal-pitch roof is one in which the slopes of the main roof are different from those of the intersecting gables or hips. Unequal pitches occur when the ridges of the roof are of the same height and the spans are different, or when the spans are alike and the ridges are at different heights.

METHODS OF FRAMING ROOFS

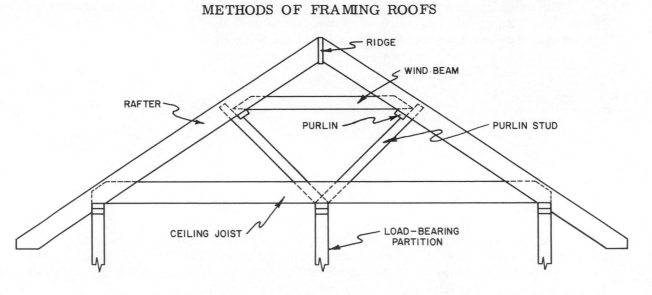

FIG. 14-1 ROOF BRACING DETAIL

The frame of any roof is made up of numerous rectangular timbers called rafters. These are inclined in pairs, their lower ends resting on the wall plate and their upper ends fastened together, usually by means of a ridgeboard which extends the full length of the house. In joist-and-rafter construction the ceiling joists extend from the exterior walls to a load-bearing partition, which also runs the length of the house. The ceiling joists act as a tie and prevent the rafters from spreading. If the rafters are braced, the length of the span can be increased. Joist-and-rafter construction is usually limited to roofs with slopes greater than five inches of rise in every twelve inches of run. Lesser slopes are possible when a vertical support is used at the peak.

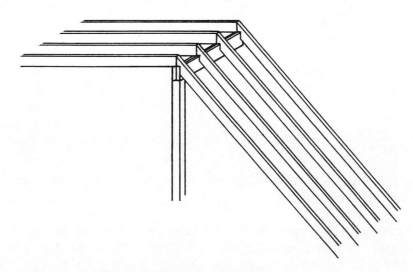

FIG. 14-2 RIDGE-SUPPORT METHOD

106

A ridge beam is sometimes used with roof slopes of less than 5″ of rise to 12″ of run. The beam is used as a ridge support with the rafters resting upon it. The beam is supported by posts or by a load-bearing partition. Ceiling finish material is applied to the underside of the rafters after the insulation, vapor barrier, and utilities have been installed (Fig. 14-2).

When plank-and-beam construction is used, the roof may be framed with transverse beams running at right angles from a ridge beam to the exterior walls. The beams may be placed up to eight feet apart; therefore, two-inch tongue-and-groove material must be used for the roof sheathing.

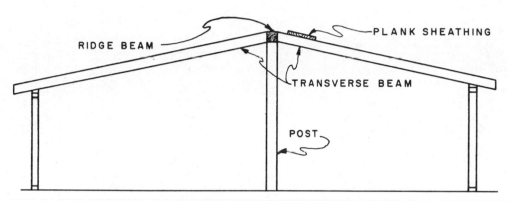

FIG. 14 –3 TRANSVERSE-BEAM METHOD

Another roof-framing method for plank and-beam-construction is with longitudinal beams. Usually three parallel beams are used, running the length of the house. The beams are supported by the end walls and posts or partitions. The two-inch tongue-and-groove roof sheathing is applied extending from the ridge beam to the side wall plates. The insulation is placed on the top of the roof sheathing, and the underside decorated as a finished ceiling.

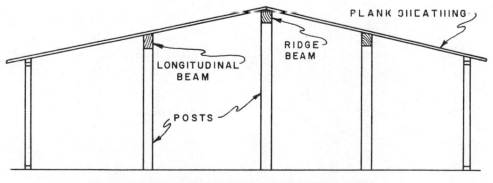

FIG. 14 – 4 LONGITUDINAL-BEAM METHOD

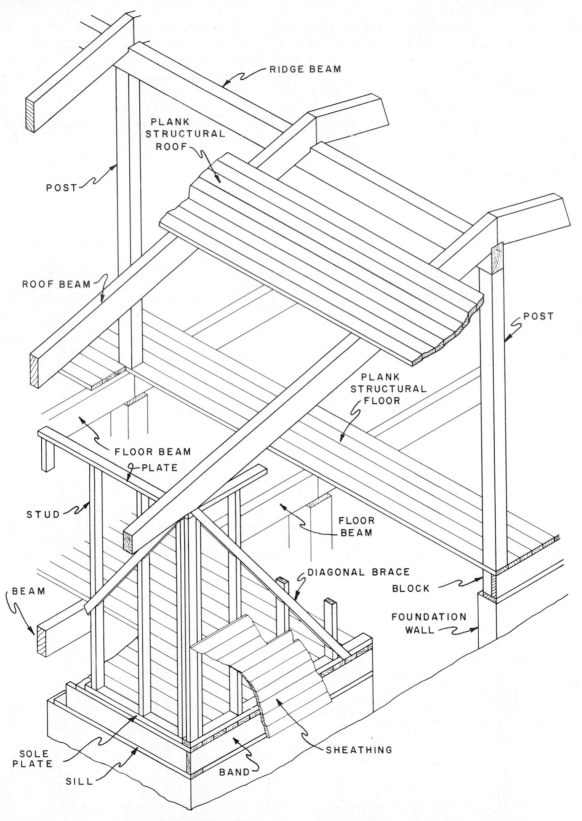

RIDGE BEAM

PLANK
STRUCTURAL
ROOF

POST

ROOF BEAM

PLANK
STRUCTURAL
FLOOR

POST

FLOOR BEAM
PLATE

STUD

FLOOR
BEAM

DIAGONAL BRACE

BLOCK

BEAM

FOUNDATION
WALL

SOLE
PLATE

SHEATHING

SILL

BAND

FIG. 14-5   PLANK-AND-BEAM FRAMING FOR ONE-STORY HOUSE

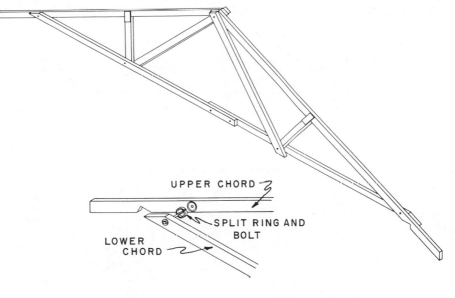

FIG. 14—6   TRUSSED RAFTER ASSEMBLED WITH
TIMBER CONNECTORS AND NAILS

Lightweight roof trusses which rest on the exterior walls and span the entire width of the house have proven suitable for homes. The trusses consist of a top chord and a bottom chord connected by vertical and diagonal braces. The members are fastened together by bolts, connectors, or gusset plates. The trusses are fastened to the exterior walls and need no center support; therefore, the partitions can be nonbearing and can be installed or relocated easily. The design of the trusses should conform with sound, accepted engineering practice.

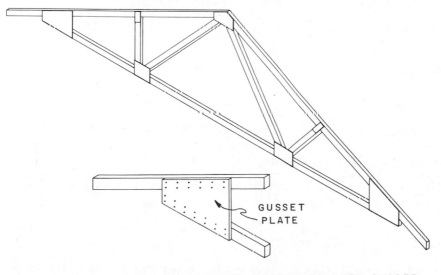

FIG. 14—7   TRUSSED RAFTER ASSEMBLED WITH GUSSET
PLATES, GLUE, AND NAILS

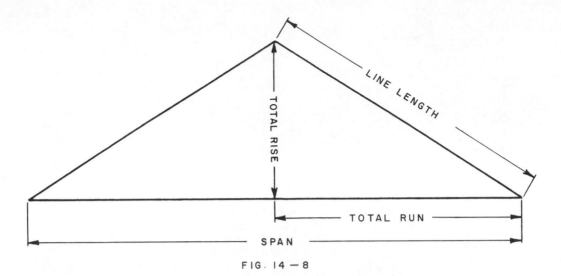

FIG. 14 — 8

TERMS USED IN ROOF FRAMING

The span of a roof is the level distance between the outer faces of the roof plates on opposite sides of the building.

A rafter is one of the sloping members of a roof which supports the roof covering in the same manner as the joists support the flooring.

The ridge is the uppermost horizontal line of a sloped surface.

The total run of a rafter is one-half the span distance. This is the level distance over which the rafter rises.

The total rise is the vertical distance from the top of the plate to the top of the ridge. Multiply the span by the pitch to find the total rise.

The line length of a rafter is the hypotenuse of a right triangle whose base is the total run and whose altitude is the total rise.

The pitch of a rafter is the amount of slope of a roof. The pitch may be given in many different ways. In terms of proportion it is found by dividing the total rise by the span.

The unit of run is the unit of measurement, 12 inches or 1 foot. As the run measurements are always taken in the level plane, the unit selected is 12 inches or 1 foot as a basis for all roof framing.

The rise in inches is also known as the rise per foot of run and means that the roof rises that much for every foot of run of the rafter.

The cut of the roof is stated as the rise in inches and the unit of run. For example, 4 and 12, 7 and 12, etc.

A plumb line is any vertical line on the rafter when it is in its proper position.

A level line is any horizontal line on the rafter when it is in its proper position.

110

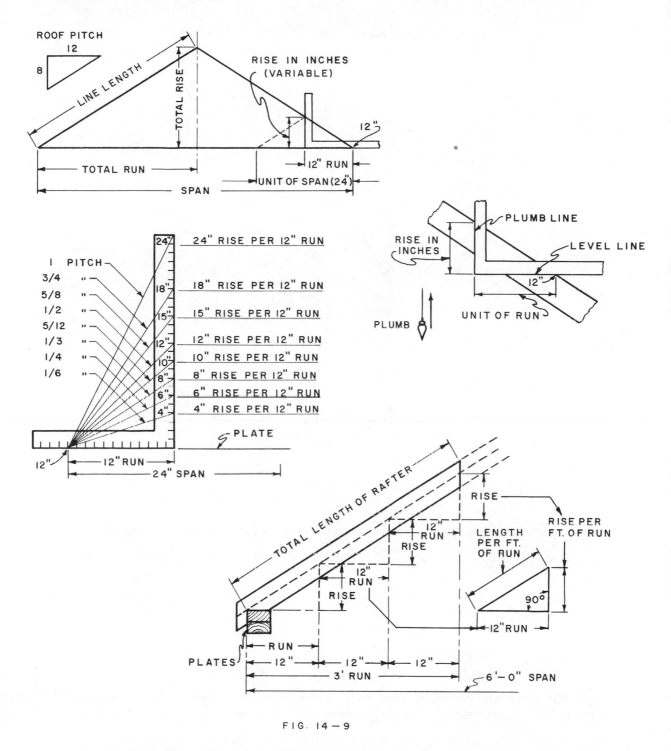

FIG. 14—9

A <u>common rafter</u> runs at a right angle from the ridge to the plate. This rafter is one of the main structural supports of the roof. The common rafter is used as a basis for the layout of the other kinds of rafters in a roof.

A <u>jack rafter</u> is one which extends from a ridge to a valley or from a plate to a hip rafter.

STRUCTURAL MEMBERS OF ROOFS

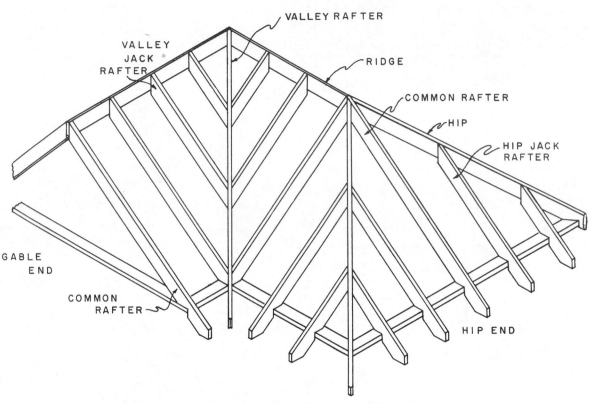

FIG. 14 — 10   HIP-AND-VALLEY ROOF

A hip rafter is one which connects two intersecting slopes of a roof at the hip.

A valley rafter is one which connects two intersecting slopes of a roof at the valley.

A cripple rafter is a short rafter which extends from a hip to a valley but does not reach the plate or ridge. It is generally made from the same stock as the common rafters if it is over 5 ft. long, but may be made from narrower lumber if it is shorter.

A collar beam holds the rafters from spreading apart. It is used where the span of the roof is excessive or where the roofing material is heavy.

The overhang of the rafters extends beyond the edge of the plate, providing support for cornice and for drain gutters. This section of the rafter is called the tail.

A tail rafter is often ornamented and is nailed on the side of a common rafter. It is used where the tail of the rafter is exposed to view, such as in an open cornice. Brackets supporting the tails of rafters are called lookout brackets.

The ridgeboard is placed between the uppermost ends of the rafters, which are then nailed together on opposite sides of this board. It serves to tie the rafters together and makes the erection of the roof easier.

FACTORS TO BE CONSIDERED IN PLANNING A ROOF

In general, all roofs which are to be shingled should have a rise of at least 8″ to the foot. Roofs of less rise should be covered with a continuous sheet of material such as metal, canvas, or composition covering.

Where locale dictates, provision for snow load should be made when a roof with a pitch of less than 20° is being built. In the central part of New York State, this allowance should be from 20 to 25 pounds per square foot.

The weights of roofing materials must also be considered. The following weights include roof boards, rafters, and the roofing material.

Asbestos cement shingles	8 to 12 lbs. per sq. ft.	Slate and felt	10 to 15 lbs. per sq. ft.
		Tar and gravel	10 to 14 lbs. per sq. ft.
Asphalt shingles	4 to 6 lbs. per sq. ft.	Corrugated iron	
Wood shingles	6 to 10 lbs. per sq. ft.	on rafters	4 to 6 lbs. per sq. ft.
Tin and felt	6 to 8 lbs. per sq. ft.	Tile	15 to 22 lbs. per sq. ft.

The above weights take into account the pitch of the roof. For example, the slate-and-felt roof is figured at 10 to 15 lbs. per sq. ft. and a slate roof must have at least a 1/3 pitch. Tar and gravel, which is much lighter than slate, is figured at 14 lbs. per sq. ft. because tar and gravel are used on a flat roof. The flatter the pitch of the roof, the greater the danger of sagging from snow loads. It can readily be seen that the roofing material has a direct bearing on the roof structure.

MAXIMUM SPAN FOR RAFTERS — No. 1 COMMON					
Roof Load of 30 lbs. per Sq. Ft. Uniformly Distributed for Slopes of 20 degrees or more (Note: A 20° slope is approximately a 4 1/4″ rise in a 12″ run)					
Standard Lumber Sizes		Distance on Center	Maximum Clearance Span — Plate to Ridge		
Nominal Size	Actual Size		South. Pine & Douglas Fir	Western Hemlock	Spruce
2″ × 4″	1 1/2″ × 3 1/2″	16″	7′-2″	6′-8″	6′-2″
		24″	5′-10″	5′-6″	5′-0″
2″ × 6″	1 1/2″ × 5 1/2″	16″	14′-0″	13′-8″	11′-2″
		24″	11′-6″	11′-6″	9′-0″
2″ × 8″	1 1/2″ × 7 1/4″	16″	18′-4″	17′-8″	16′-4″
		24″	15′-8″	15′-8″	13′-4″
2″ × 10″	1 1/2″ × 9 1/4″	16″	21′-10″	21′-2″	20′-4″
		24″	19′-8″	19′-0″	16′-10″

Dead load figured to include weight of rafters, roof sheathing, and roofing. For heavier roof covering, use rafters next size larger.

NOTE: All sizes given are based on use of No. 1 Dimension lumber. In figuring No. 2 Dimension use the next shorter span or the next larger size.

The rafters carry the load of the roof and are similar to floor joists in the way they support a load. The material on load-carrying capacity of girders in Unit 4 might be applied in finding the sizes of hip and valley rafters. The hip or valley rafter might be considered as the girder, and the jack rafters as joists. The previous chart shows a rapid means of determining the maximum spans for common rafters.

REVIEW PROBLEMS   Unit 14

1. Name the two basic types of roof construction.

2. How much slope may a so-called flat roof have?

3. Why is a built-up roof covering of bituminous material not used on a pitched roof?

4. Describe several types of roof.

5. What is the difference between an equal-pitch and an unequal-pitch roof?

6. What is the minimum slope recommended for a joist-and-rafter roof?

7. A ridge beam is sometimes used with roof slopes of less than 5″ of rise to 12″ of run. How is the beam supported?

8. Explain the difference between a transverse- and a longitudinal-beam roof.

9. How can insulation be applied to a beamed roof?

10. How can members of a roof truss be fastened together?

11. State two advantages of a trussed-roof system.

12. Explain the terms: (1) span, (2) total run, (3) total rise, (4) line length.

13. How long is the unit of run which is the basis for all roof framing?

14. What is the cut of a roof?

15. Explain the meaning of a plumb line and a level line in terms of a rafter.

16. Describe the location of: (1) common rafter, (2) hip rafter, (3) jack rafter, (4) valley rafter, (5) cripple rafter.

17. List several factors to be considered in planning a roof.

Unit 15   COMMON RAFTERS

There are several ways of finding the lengths and cuts of roof rafters. However, the basic principles of all methods is geometric construction. Each of the common methods — the graphic, the rafter table, and the step-off — is used where it can be applied most conveniently. The architectural draftsman might find the graphic method most convenient because the work may be laid out with drafting instruments. The carpenter finds the step-off or the rafter-table method convenient because the work is laid out with a framing or steel square.

The step-off method of laying out rafters is considered the most logical for the carpenter to use, so this unit will describe that procedure. The rafter tables will also be discussed so that both methods may be used as a check against inaccurate work.

To avoid confusion, the learner should study only one system of rafter layout at a time until he is thoroughly familiar with it. These units on roof framing contain only the fundamental principles necessary for laying out the type of rafter under discussion.

Rafter layout is a somewhat complicated procedure to learn. Therefore, it is important that some definite habits be formed as to the order in which the various steps of the layout work are performed.

In laying out rafters, they should all be placed in the same relative position. The crowned edge, which should be considered the top edge, should be toward the carpenter who is laying out the rafter.

The tongue of the steel square should be held in the left hand and the body of the square in the right.

If the square is pictured as greatly enlarged, the tongue should form the vertical or top cut of the rafter, and the body, the level or seat cut.

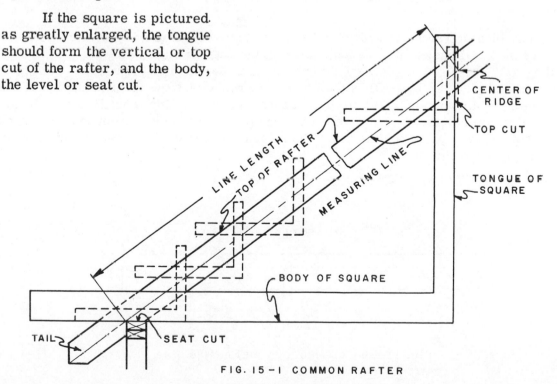

FIG. 15-1 COMMON RAFTER

FINDING THE ROUGH LENGTH OF A COMMON RAFTER

The approximate length of a common rafter may be found by representing the rise of the rafter in feet on the tongue of the square and the run in feet on the blade. The length of the diagonal between these two points is then measured. This measurement, expressed in feet, is the rough length of the rafter. If an overhang for a cornice is needed, this length should be added.

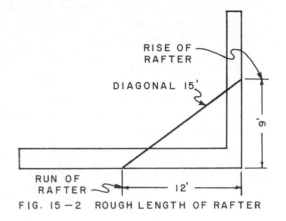

● Example: Assume that the total rise of a rafter is 9 ft. and the run is 12 ft. Locate 9 and 12 on the square and measure the diagonal. This will be found to be 15. Therefore, the rough length of the rafter will be 15 ft. plus the overhang of 1 ft. Stock 16 ft. long would have to be used.

FIG. 15—2   ROUGH LENGTH OF RAFTER

LOCATING THE MEASURING LINE ON A RAFTER

The stock of which the first rafter is made should be straight and of the correct rough length. The width and thickness may be determined as described in Unit 14. This piece is laid flat across two sawhorses and the square is placed near the right-hand end.

The inch mark on the outside edge of the body corresponding to the run of the roof (12 in. in all cases) and the inch mark on the outside edge of the tongue corresponding to the rise of the roof, should both come at the edge of the rafter. This is to be the top edge. The line AB is then drawn to represent the level of the top of the wall plate. A distance of 3 1/2 in. is measured along this line from B to C to locate the outside top corner of the plate. This should be far enough from the right-hand end of the rafter to allow for the tail. The measuring line (CD) is then gaged parallel to the edge of the rafter.

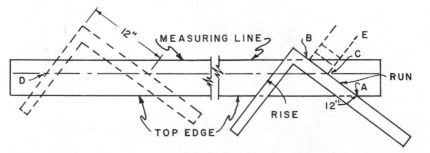

FIG. 15—3   LAYING OUT MEASURING LINE

LAYING OUT A COMMON RAFTER (STEP-OFF METHOD)

A wooden fence may be made and clamped at the proper points on the tongue and body of the square to aid in laying out a rafter. Small metal clamps may be used for this purpose.

Assume that a building is 24 ft. wide and the pitch of the rafter is to be 1/3, or 8" rise on 12" run. It will then be necessary to find the exact length of the rafter. The square is laid on the stock so the 12" mark on the outside of the body is at point C, and the 8" mark on the outside edge of the tongue is on the measuring line at E. The square is kept in this position, and the fence is adjusted so its edge lies against the top edge of the rafter. The fence is then tightened on the square. It is important that the 12" mark on the body of the square and the 8" mark on the tongue be exactly on the measuring line when the fence is against the edge of the rafter.

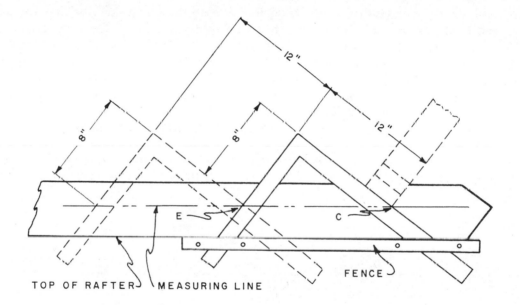

FIG. 15-4   STEP-OFF METHOD

Procedure

1. Place the fence against the rafter stock as shown, Fig. 15-4.

2. Mark along the outside edge of the body and tongue of the square. This locates point E on the measuring line.

3. Slide the square to the left until the 12" mark is over E (position 2).

4. Again, mark along the tongue and body of the square.

5. Continue this operation as many times as there are feet of run of the rafter (12 in this case), Fig. 15-5.

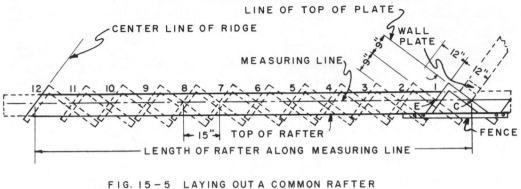

FIG. 15-5 LAYING OUT A COMMON RAFTER
(STEP-OFF METHOD)

The successive positions of the square are shown by the numbers 1 through 12. When the last position has been reached, a line drawn along the tongue across the rafter will indicate the centerline of the ridgeboard.   Each step in this whole process should be clearly marked and performed very carefully since even a slight error will greatly affect the fit and length of the rafter.

LAYOUT OF A RAFTER WHEN THE SPAN IS AN ODD NUMBER OF FEET

Assume that a building is 25 ft. wide and the pitch of the rafter is 1/3 or 8" rise to 12" run. The run of the rafter would be one-half the width of the building, or 12'-6".

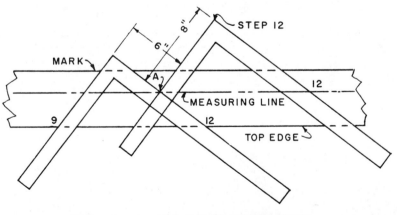

FIG. 15-6 ADDITIONAL HALF STEP

Procedure

NOTE: The layout of the rafter is very similar to the one previously described, except 12 1/2 steps are taken instead of 12. The additional 1/2 step is added for the extra 6" run of the rafter.

1. After the twelfth step has been taken and marked, Fig. 15-6, place the square on the top edge of the rafter with the numbers 8 and 12 coinciding with the edge (see Fig. 15-6).

2. Move the square until the number 6 on the outside of the body is directly over the 12th step plumb line.

3. Mark a line along the outside edge of the tongue to indicate the centerline of the ridgeboard.

ALLOWANCE FOR RIDGEBOARD

The last line marked on the rafter shows where the rafter would be cut off if there were no ridgeboard and the rafters were to be butted against each other. Since a ridgeboard is used in most cases, the rafter as previously laid out will be a little too long. It will be necessary to cut a piece off the end of the rafter equal to half the thickness of the ridgeboard.

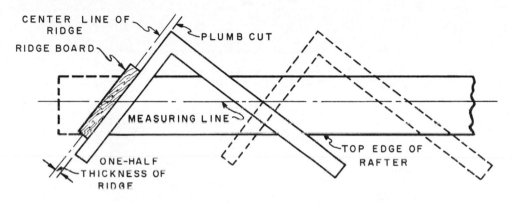

FIG. 15 — 7   ALLOWANCE FOR RIDGEBOARD

Procedure

1. Slide the square back away from the last line (position 12), keeping the fence tight against the upper edge of the rafter, Fig. 15-7.

2. When the square has been moved back half the thickness of the ridgeboard from this last line, mark the plumb cut along the edge of the tongue.

    NOTE: This line should be inside of, and parallel to, the original line marking the end of the rafter. The measurement should be taken from this original line, and at right angles to it.

BOTTOM OR SEAT CUT

The bottom or seat cut of the rafter is a combination of the level and plumb cuts. The level cut (DF) rests on the top face of the side wall plate and the plumb cut (DC) fits against the outside edge of the wall plate. The plumb cut is laid out by squaring a line from DF through point D. This line (DC) then represents the plumb cut. (See Fig. 15-8.)

TAIL OF THE RAFTER

If the rafter tail is to be of the type shown, the level and plumb cuts are laid out as before. Since the sheathing extends into the rafter notch up to the top of the plate, an allowance for the thickness of the sheathing must be made. This can be done by laying out line AB parallel to line CD and the thickness of the sheathing away from CD. The line of the level cut at D is then extended to meet AB. When the rafter is cut on these lines, it will fit over the plate and sheathing.

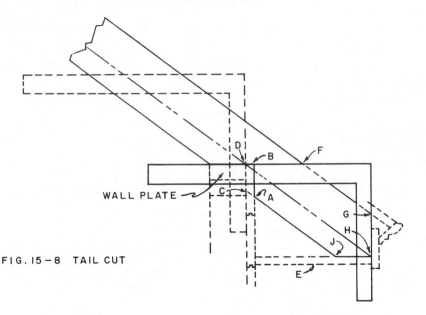

FIG. 15—8 TAIL CUT

Procedure

1. Assuming that a **12″** piece is to be used at E, continue the line of the level cut through B to F on the top edge of the rafter.

2. Lay the body of the square along this line with the 12″ mark of the outside edge directly over point B.

3. Mark a line GH along the outer edge of the tongue across the rafter. The point where this line meets the measuring line at H represents the tip of the rafter.

4. Square the line HJ across the rafter from line GH to locate J.

RAFTER TABLES

Many of the so-called steel squares manufactured today have rafter tables stamped on them. They are properly referred to as a framing square because of the various scales and tables which are especially adapted for use in the framing of the numerous members of a frame house. The tables on the different makes of framing squares are not the same and not all steel squares have the framing tables stamped on them. Further information on the framing square may be found in the book, Hand Woodworking Tools, of this series. A booklet giving instructions on the use of these tables may be secured from the manufacturer of the square. If the principles involved in roof framing are understood, the use of the tables can be quickly learned.

LENGTH	COMMON	RAFTERS	PER FOOT	RUN	21 63	20 81	20 00	19 21	18 44	17 69	16 97	16 28	15 62	15 00	14 42
"	HIP OR	VALLEY	"	"	24 74	24 02	23 32	22 65	22 00	21 38	20 78	20 22	19 70	19 21	18 76
DIFF	IN LENGTH	OF JACKS	16 INCHES	CENTERS	28 84	27 74	26 66	25 61	24 585	23 588	22 625	21 704	20 83	20	19 23
"	"		2 FEET	"	43 27	41 62	40	38 42	36 88	35 38	33 94	32 56	31 24	30	28 84
SIDE	CUT	OF	JACKS	USE	6 11/16	6 4/16	7 3/16	7 1/2	7 7/16	8 1/8	8 1/2	8 7/8	9 1/4	9 5/8	10
"	"	HIP OR	VALLEY	"	8 1/4	8 1/2	8 3/4	9 1/16	9 3/8	9 5/8	9 7/8	10 1/8	10 3/8	10 5/8	10 7/8

FIG. 15-9   FRAMING TABLE

After the rise and run of the rafter have been determined, the length of the rafter along the measuring line may be found from the rafter table. The inch marks on the outside edge of the square are used to denote rise per foot of run. For example, the figure 8 means 8 in. of rise per foot of run. Directly below each of these figures is the length of main rafters per foot of run, 14.42 in this case. If the run of the rafter is 10 ft., the 14.42 in. should be multiplied by 10. This gives 144.20 in. or 12.01 ft. as the length of the rafter. The other steps in laying out a rafter are the same as described under the step-off method.

TOOLS AND EQUIPMENT FOR FRAMING COMMON RAFTERS

Hammer	Straightedge	Rule
Crosscut and ripsaw	Steel square	
Spirit level	Portable electric saw	

How to Lay Out and Cut Common Rafters

1. Check the actual width of the building at the attic plate line to find out if the rafters at the seat cut are to butt against the outside of the sheathing or directly against the plate.

2. Lay out and cut one rafter to be used as a pattern.

3. Nail a piece of sheathing about 6 in. long near each end on the top edge of this rafter as shown. This will make it easier to line up the pattern with the other rafters when they are ready to be marked.

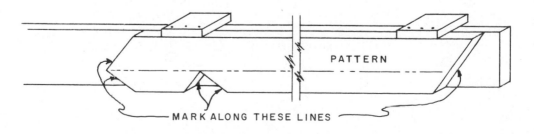

MARK ALONG THESE LINES

FIG. 15-10   RAFTER PATTERN

4. Select the required number of rafters and pile them with the crowned edges all facing one way. Do not use any rafters which have serious defects.

5. Lay the pattern on each rafter and mark the cuts.

6. Make the ridge, seat, and tail cuts. These cuts should be square with the side of the rafter. A ripsaw or portable electric saw is often used to make these cuts.

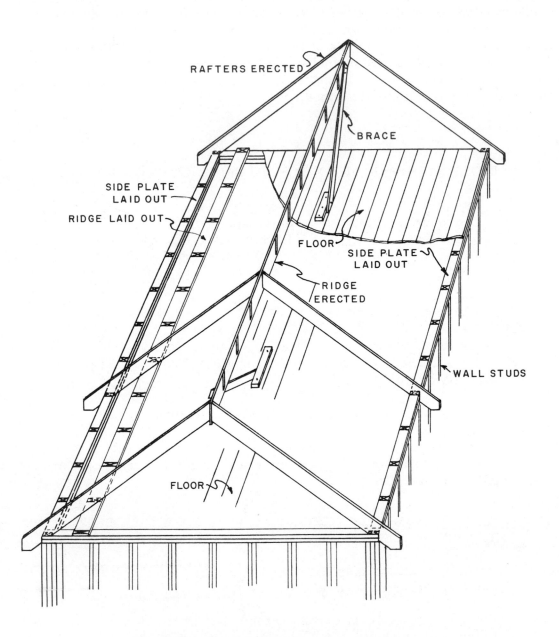

FIG. 15—11 PLATE AND RIDGE LAYOUT

How to Space Rafters on the Plates and Ridgeboard

1. Secure a sufficient number of boards to make up the length of the ridgeboard. These boards should be as wide as the plumb cut is long.

2. Lay the boards along the plate on one side of the building.

3. Mark the spacing for the rafters on the plate and on the ridgeboard. Use the same method as in marking the ribbon for the balloon frame. The joints of the ridgeboard should come halfway on a rafter. Let the last rafter space come where it will.

4. Cut the last piece of ridgeboard off flush with the outside of the plate at the end of the building.

5. Place the ridgeboard at the side of the plate on the opposite side of the building and mark this plate the same as the ridgeboard.

How to Erect and Secure Rafters on a Gable Roof

1. Build a scaffold about 10 ft. long for erecting the rafters. The top of the scaffold should be 5 ft. from the underside of the ridge and should be so erected that it can be moved along the attic floor.

2. Place the ridgeboard on the scaffold with the ends pointing the same way as when they were laid out.

3. Select four of the straightest rafters for gable ends.

   NOTE: When the ridgeboard is made up of several lengths, a group of rafters is nailed on the first length of ridgeboard and to the plate. This section is then braced. The scaffold is then moved along the floor and the next section of rafters is raised.

4. Nail the first length of ridgeboard to the end rafter. The end of the ridge should be flush with the outside of the rafter. The top of the rafter should be on the spacing marks and even with the top of the ridgeboard. Nail through the side of the ridgeboard and into the ridge cut of the rafter with three 8d nails.

5. Nail another rafter on the same side of the ridgeboard but about five rafter spacings away.

6. Raise these rafters until the seat cut fits at the plate and toenail them in place at the bottom marks with 16d nails.

7. Place the opposite rafters on the other side of the ridgeboard and nail them to the plate at the seat cut. Then raise or lower the ridgeboard until it is even with the top of these rafters. Nail the rafters in position through the opposite side of the ridgeboard or toenail them.

8. Plumb and brace this section by nailing one end of a 2 × 4 stay to the ridge-board at the top of the gable and the other end to a block spiked to the attic floor. This block should be spiked into at least two joists.

9. Nail the rest of the rafters of this section in place. Erect first one rafter and then the opposite one. Nail the bottom end of the rafter first to provide a straighter ridge.

   NOTE: One of the rafters at the joint in the ridge should be tacked a little to one side until the rest of the ridge is in place.

10. Move the scaffold along the floor so that the other section of the ridge may be raised in a similar manner. It is assumed that there are only two pieces of ridgeboard.

11. Nail the outer end of the second piece of ridgeboard to the outside or gable rafter, keeping the end flush with the face of the rafter.

12. Nail one rafter near the center of this piece of ridgeboard.

13. Raise these rafters and place the inside end of the ridgeboard tight against the other piece of ridgeboard. Nail the second piece of ridgeboard to the rafter that centers on the joint.

14. Raise the other gable rafter and nail the bottom and then the top.

15. Move the rafter which was tacked to one side of the end of the first piece of ridgeboard and nail it in place over the joint in the ridgeboard.

16. Erect the rest of the rafters as described in Step 9.

17. Fasten a brace to the top of the gable as in Step 8.

   NOTE: Check the rafters of both gables to see if they are plumb. If one gable is plumb and the other is not, the ridgeboard was laid out wrong or the joint was not drawn together. Correct this before proceeding.

How to Lay Out, Cut and Install Gable Studding

1. Square a line across the plate directly below the center of the gable.

2. If a window is to be installed in the gable, measure one-half of the opening size on each side of the centerline and make a mark for the first stud.

3. Starting at this mark, lay out the stud spacing 16 in. o.c. to the outside of the building.

4. Stand a 2 × 4 stud upright on the first mark. Place it against the side of the gable rafter and plumb it. See first stud (Fig. 15-12).

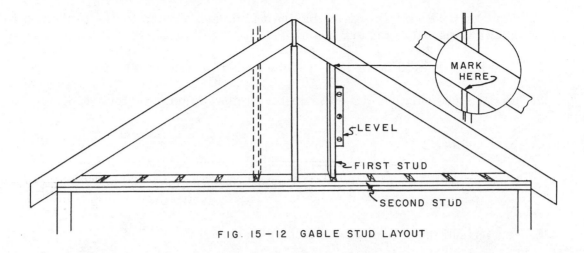

FIG. 15 – 12   GABLE STUD LAYOUT

5. Mark across the edge of the stud at the underside of the rafter (enlarged view).

6. Stand a 2 × 4 on the second stud mark.

7. Plumb and mark it as described for the first stud.

8. Measure the difference in length between these two studs. Each remaining stud will be this distance shorter than the preceding one.

9. Mark all the remaining studs.

10. Cut two studs of each length, one for each side of the gable.

11. Nail the studs in position flush with the outside of the rafters and plate.

12. Frame in the window sills and headers as described in Unit 11.

REVIEW PROBLEMS   Unit 15

1. List three methods of finding the lengths and cuts of roof rafters.

2. State two methods commonly used by the carpenter to find the lengths and cuts of roof rafters.

3. When laying out a rafter, how should the rafter stock be placed in relation to the worker?

4. How would you find the approximate length of a common rafter with the aid of a framing square?

5. Explain the method used to find the location of the measuring line on a rafter.

6. What precaution should be taken when the fence is fastened to the framing square?

7. What will the last line drawn along the tongue across the rafter indicate, when laying out a rafter?

8. How much should be deducted from each rafter for a ridgeboard?

9. Do all steel squares have the rafter tables stamped on them?

10. Are the rafter tables and scales the same on different manufacturers' squares?

11. What is the purpose of fastening a short piece of sheathing near each end of the rafter pattern?

12. How should the crown edge of rafters be placed?

13. Where should the joint in a ridgeboard be located?

14. Should the straightest rafters be selected for the gable ends of a roof?

15. How should the rafters be fastened to the ridgeboard?

16. What precaution should be taken at the gable ends of the roof after the rafters are in place?

17. Describe the method of laying out the location of the gable studs.

18. List the tools necessary to cut and install common rafters.

Unit 16   HIP AND VALLEY RAFTERS

The method of stepping off the lengths and cuts of hip or valley rafters is very similar to that of the common rafter, except that the unit of run of the hip or valley rafter is greater than the unit of run (12 inches) of the common rafter. The common rafter meets the ridgeboard and the plate at a 90° angle, whereas a hip or valley rafter meets these members at an angle of 45°. This accounts for the greater unit of run of the hip or valley rafter. This also makes it necessary to cut the common rafters where they meet the hip or valley.

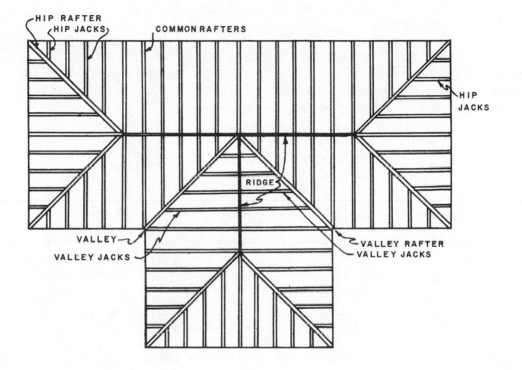

FIG. 16—1

A roof plan may be drawn to scale on paper, or chalk lines indicating the location of the rafters may be made on the attic floor. This will show the location, number, and run of the various rafters.

The six hip rafters are shown at the ends of the main roof and at the end of the projection. The hip-jack rafters extend from the hip down to the plate. The valley rafter is shown at the intersection of the main roof and the side roof. The valley jack rafters extend from this valley rafter up to the ridge.

The pitch of the hip and that of the valley are exactly the same. The pitch of the common rafter of the main roof is the same as that of the common rafter of the projection. This type of roof is called a hip-and-valley roof of equal pitch.

127

RUN OF HIP AND VALLEY RAFTERS

The common rafter extends from the plate to the ridge with **12** inches used as the unit of run. This unit of run is shown in the sketch as two sides of the square ABCD.

The hip rafter must also extend from the plate to the ridge, but at an angle of 45° with the plate. See line BD. Therefore, the diagonal of the 12″ square is used as the unit of run for the hip rafter. This is 16.97″, or approximately 17″. The length of a hip or valley rafter for a roof of equal pitch is found by using 17″ on the body of the square as the unit of run instead of 12″ as for the common rafter.

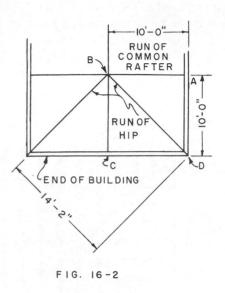

FIG. 16-2

If the rise per foot of run of the common rafter is 8″, the length of the hip may be found by following the same step-off method as laying out a common rafter, except that the fence should be set on the body of the square at 17″ and on the tongue at 8″. The number of steps to be taken will be equal to half the span of the building in feet, the same as for a common rafter on the same roof.

The method of locating the intersection of the hip rafters with the ridgeboard is shown at B. Notice that the distance AB is equal to BC, both distances representing the run of the common rafters. The two squares are equal in size. The distance DB is the diagonal of the square ABCD and represents the run of the hip rafter.

LAYOUT OF A HIP RAFTER

Assume that a hip rafter is to be laid out for the roof shown in the previous sketch. The rise of the common rafter is 8″ per foot of run. The proper width and thickness of the hip rafter may be selected by referring to the table of rafter sizes and strengths in Unit 14.

The length of the hip rafter may be found by following the method explained in Unit 15.

The measuring line is gaged the same distance from the top of the hip rafter as from the top of the common rafter.

The length of the hip rafter is found by setting the fence on the square at 17″ on the body and 8″ on the tongue and stepping along the rafter for 10 steps.

The seat cut is found in the same way as for the common rafter, except that 17″ is used on the body of the square and 8″ on the tongue instead of 12″ and 9″ as in Unit 15.

RIDGE CUT FOR HIP RAFTER

An enlarged plan view of the ridge at the intersection of the hip and common rafters is shown in Fig. 16-3. Notice that the common rafters 1 and 2 butt against two sides of the ridgeboard. Rafter 3 butts against the end of the ridge and acts as a 1 1/2″ ridge between the two hip rafters. The hip rafters fit into the angles made by the three common rafters. The ridge cut at the top of each hip will, therefore, consist of several bevel cuts.

A. Allowance for the Thickness of the Ridge

The allowance to be taken off the ridge cut at the top of the hip for the thickness of the ridgeboard is somewhat different from that of the common rafter. To show this allowance more clearly, Fig. 16-3, centerlines have been drawn on all the members of the roof at this point. The meeting point of these centerlines is located at B. The distance from point B to the intersection of the hip and common rafter, point C, changes as slope of the roof is changed and as the thickness of the ridge is changed. In this case, where a 3/4″ ridge is used on a 1/3-pitch roof, the distance BC is approximately 1 1/4″. If, however, a ridgeboard 2″ thick were used for a 1/6-pitch roof, the distance BC would be 1 3/8″. If a 2″ ridge were used for a 1/2-pitch roof, the distance BC would be 1 3/4″. Therefore, in ordinary-pitch roofs, 1 1/2″ may be used as the average distance BC.

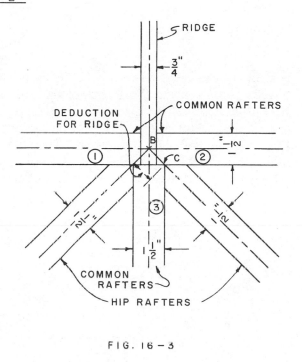

FIG. 16-3

The distance BC is marked back from the top end of the hip rafter the same way the distance was taken off the common rafter except that the figure on the tongue of the square will be 8 and on the body 17. The mark is then transferred from the side of the hip rafter to the centerline on the top edge as at C.

Fig. 16-4 shows another method of construction at the intersection of the hip rafters and ridgeboard. This shows a single cheek cut on the hip rafter and a 3/4″ ridgeboard. The deduction for the ridgeboard is also shown.

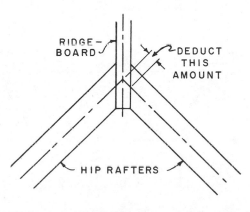

FIG. 16-4

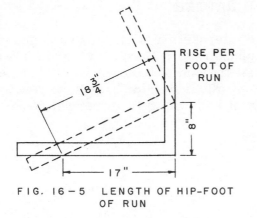

FIG. 16-5   LENGTH OF HIP-FOOT
OF RUN

B. Bevel or Cheek Cut

Before marking out the cheek cuts, it is necessary to find the length of the hip per foot of run. This may be found by taking the rise of the roof per foot of run (8″ in this case) on the tongue of the square, and the unit of run of the hip (always 17″) on the body of the square and then measuring the diagonal. In this case it is approximately 18 3/4″, Fig. 16-5.

To obtain the angle of the double cheek cut, lay the square on the top edge of the hip rafter, Fig. 16-6. The figure on the body representing the run of the hip (17″) should lie on the centerline, and the figure on the tongue representing the length of the hip per foot of run (18 3/4″ in this case) should lie on point C. Since the tongue of the square is only 16″ long, it will be necessary to use figures which are just half of those just given. Therefore, 8 1/2 would be used on the body and 9 3/8 on the tongue. The line CD should then be marked along the outside edge of the tongue of the square. A line should be squared across the edge of the rafter from D to E. A line connecting E and C should then be drawn.

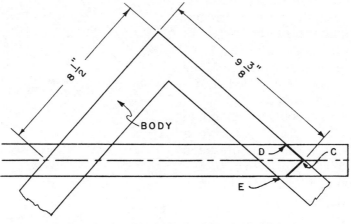

FIG. 16-6   DOUBLE-CHEEK CUT OF HIP

C. Plumb Cut

The plumb cut is marked from D and E down both sides of the rafter. In this case, this cut may be obtained by taking 8″ on the tongue of the square and 17″ on the body and marking along the outside of the tongue, Fig. 16-7.

D. Tail Cuts

The tail cuts for the cornice at the bottom of a hip or valley rafter are found in a similar manner to those of the common rafter. However, the figures 8 and 17 on the square should be used as shown.

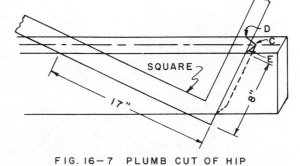

FIG. 16-7   PLUMB CUT OF HIP

BACKING OFF THE HIP RAFTER

Since the top edges of a hip rafter are at the intersection of the side and end slopes of the roof, they project above these slopes as shown, Fig. 16-8. The edges must be beveled off to provide an even plane from the top of the jack rafters to the centerline of the hip. The roof boards will then lie flat at the hip.

The bevel cuts shown by dotted lines on each side of the top of the hip rafter are rather short and it is difficult to lay them out with the steel square. However, a pattern may be laid out to show the cuts, Fig. 16-9.

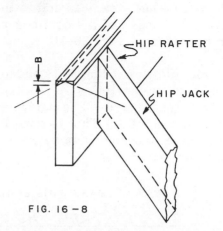

FIG. 16-8

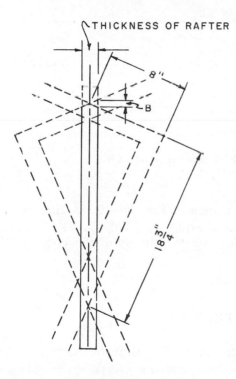

FIG. 16-9 BEVEL FOR BACKING HIP RAFTER

A centerline is gaged along a piece of stock about 3/4" thick, 1 1/2" wide and 30" long. The length of the hip per foot of run (18 3/4" in this case) is determined as explained under the bevel or cheek cuts. The square is then placed across the pattern so that 18 3/4" on the body and the rise of the hip rafter (8") on the tongue both come on the centerline. A line is marked along the tongue. The square is reversed to get the same angle on the opposite side of the centerline of the pattern.

The pattern is cut along the bevel lines and used as a templet to mark the hip section. A centerline is then gaged along the entire length of the top of the hip. Lines are also gaged along the sides of the hip. These marks show the portion of the rafter to be backed off and also mark the location of the tops of the jack rafters.

Some carpenters prefer the method of dropping the hip rafter. This is done by cutting the distance B, Fig. 16-8, off the seat cut of the hip, thus avoiding backing off the hip rafter.

LAYING OUT THE LENGTH OF THE HIP RAFTER IF THE SPAN OF THE COMMON RAFTER IS IN ODD FEET

Assume that it is necessary to lay out a hip rafter for a building 21' wide. The rafter has an 8" rise to the foot. In this case, it will be necessary to take 10 and a fraction steps.

It has already been pointed out that the unit of run of a hip rafter is 17. There-fore, it will be necessary to add to the 10 regular steps a certain proportion of 17. Since the extra span of the common rafter is 6″, it will be necessary to add 6/12 of 17″ or 8 1/2″, see Fig. 16-10.

This additional length is added to the hip rafter by taking an 11th step in the same manner as the 10th step was taken. A line is marked along the outside edge of the body of the square. The half step length of 8 1/2″ is then measured along this line from the point A. This is done by placing the square so the outside edge of the body lies along the line made by the 11th step and so the figure 8 1/2 on the square is directly over the point A.

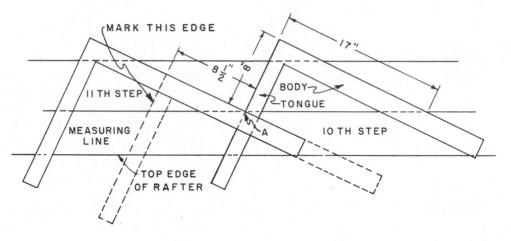

FIG. 16 – 10   ADDITIONAL LENGTH ON HIP RAFTER

This location of the square is shown by dotted lines in Fig. 16-10. A line is then marked along the outside edge of the tongue of the square. This gives the plumb or ridge cut of the hip rafter. The deduction for the thickness of the ridge is the same as previously described.

HIP JACK RAFTERS

Hip jack rafters are those rafters that extend from the plate to the hip and fill in the triangular spaces between the common rafters and the hip rafters. They fit against the plate in the same manner as the common rafters. The plumb cut of a hip jack rafter is the same as the plumb cut of a common rafter except that the cut must be beveled on the hip rafter. This additional angle must be cut because the jack rafter joins the hip rafter at an angle, whereas the common rafter joins the ridge with a square cut. This angle cut on the hip jack is called the cheek or face cut.

A.   Length of a Hip Jack Rafter

The method of finding jack rafter lengths by stepping off is similar to that used for a common rafter, explained in Unit 15.

The common rafter should first be laid out, cut, and used as a pattern, Fig. 16-11. The length of the first jack rafter is equal to the distance from point A at the seat cut along the measuring line of the common rafter to the second step taken with the square, as shown by the mark B. The length of the second jack is equal to the distance from point A to the fourth step taken with the square, or point C. The length of the third jack is equal to the distance from point A to the sixth step taken with the square or point D. The length of the fourth jack is equal to the distance from point A to the

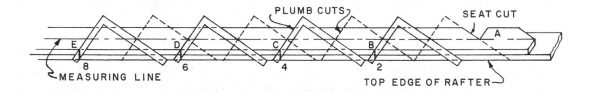

FIG. 16-11   DIFFERENCE IN LENGTHS OF JACKS

eighth step or point E. The length of any additional jack rafters is found in the same manner. There will be several jack rafters of the same size in any one roof. This method gives the lengths of jacks that are spaced 24″ o.c., which is the commonest spacing of rafters for a roof of this type.

The common rafter that is to be used for the pattern should be marked across the top edge at the points where the plumb cuts meet the top edge. See Steps 2, 4, 6, and 8. The first or shortest jack rafter is laid out by placing the common rafter pattern on the jack rafter, holding the top edges of both rafters flush. The tail cut and seat cut on the jack rafter should then be marked according to the pattern. The point marked 2 on the pattern should be squared across to top edge of the jack rafter.

B.   Cheek or Side Cut of the Jack Rafter

Fig. 16-12 shows the top end of a hip jack rafter. The mark 2 corresponds to mark 2 in the previous figure. The cheek cut is obtained by taking the unit of run of the common rafter (12″) on the body of the square, and the length of the common rafter per ft. of run, 14.42″ (14 3/8″) on the tongue. A line is marked on the outside edge of the tongue as shown. This line indicates the cheek cut. It is important to remember that the heel or short point of the bevel is toward the tail of the rafter.

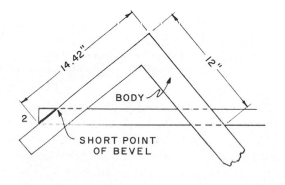

FIG. 16-12 SIDE CUT OF EACH RAFTER

C. Plumb Cut

The plumb cut is marked across each face of the rafter from each end of the line just drawn for the cheek cut. This plumb cut is marked in the same manner as that of the common rafter, using the figures 8 and 12 on the square.

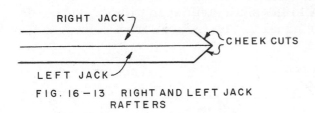

FIG. 16 – 13   RIGHT AND LEFT JACK RAFTERS

This layout is for the shortest jack on the left side of the hip rafter. A jack for the right side of the hip rafter must also be laid out. This jack is measured and laid out the same as the left jack except that the bevel or cheek cut goes the opposite way. The remaining right-side jacks are laid out as described. A corresponding left jack must be provided for each right jack.

USING THE RAFTER TABLE TO CHECK THE LENGTHS AND CUTS OF HIP RAFTERS

Finding the length of a hip rafter by means of the rafter table is similar to finding the length of a common rafter by this method. See Unit 15. A Stanley square No. R-100 is used in this case. The row of figures on the square marked Length of Hip or Valley Rafter per Foot of Run should be used.

Under the figure 8 (the rise in this case) on the edge of the square will be found the figure 18.76. This figure is then multiplied by the span of the common rafter, 10', for example, to find the length of the hip along the measuring line: $10 \times 18.76 = 187.60''$ or 15'-7 1/2". The ridge and seat cuts may be found by holding the square as in stepping off the rafter.

The side or cheek cuts for the hip rafters are found on the sixth row on the body of the square. Under 8 will be found the figure 10 7/8. This means that 10 7/8" should be taken on the body and 12" on the tongue. A line is then marked along the tongue to get the side cut.

USING THE RAFTER TABLE TO CHECK THE LENGTHS AND CUTS OF HIP JACK RAFTERS

The third line of the rafter table on the square shows the difference in the lengths of jack rafters spaced 16" on centers, and the fourth line shows the difference in lengths of jack rafters spaced 24" on centers. Since an 8" rise per foot and a 24" spacing are assumed, follow down under 8 to the fourth line. The figure 28.84 at this point means that the longest jack will be 28.84" (2'-4 7/8") shorter than the common rafter, the distance being measured along the measuring line. Each successive jack rafter will be 2'-4 7/8" shorter than the one before.

The figures for the side or cheek cuts of the jack rafters are in the fifth row on the square. Under 8, the figure 10 will be found. This means that 10 is to be used on the body and 12 on the tongue. The mark is made along the tongue for the side cut.

Valley Rafters

A valley rafter, as well as a hip rafter, extends from the plate to the ridge. The valley rafter forms the intersection between the main and projection parts of the roof. The valley rafter is about 1″ longer than the hip because it is framed against the 3/4″ ridgeboard, whereas the hip is framed against the 1 1/2″ common rafter. For hip rafters, an allowance of 1 1/2″ is deducted from the stepped-off length. For valleys, only 1/2″ is deducted for the thickness of the 3/4″ ridgeboard.

Valley Jack Rafters

The valley jacks extend from the ridgeboard down to the valley rafter. Their lengths are found in the same manner as those of the hip jacks. The ridge cut of the common rafter is used for the ridge cut of the valley jack and the cheek and plumb cut of the hip jack is used for the bottom cut of the valley jack. These jacks are also laid out in pairs.

FIG. 16—14 VALLEY JACK RAFTER

TOOLS AND EQUIPMENT FOR FRAMING HIP AND VALLEY RAFTERS

Portable electric saw	Ripsaw
Sliding T-bevel	Spirit level
Crosscut saw	Chalk line
Rule	Steel tape
Framing square	Hand axe

How to Lay Out and Cut the Rafters

NOTE: Assume that a building is to be 20′ wide and 40′ long as in the example shown in this unit.

1. Determine the proper size and number of common, valley and jack rafters.

2. Select straight, sound stock for all the roof members.

3. Lay out the common rafters as described in Unit 15.

4. Cut the common rafters as explained in Unit 15.

5. Lay out the hip rafters as explained.

6. Cut the hip rafters.

7. Determine the amount to be backed off the hip rafters as described.

8. Back off these rafters to the proper shape with a plane.

   NOTE: The hip rafters may be erected before backing them off. The lines on the sides of these rafters may then be used as guides for the location of the tops of the jack rafters. When the jacks are in place, the hip rafters may be backed off with a hand axe.

9. Lay out the jack rafters in pairs as described.

10. Cut the jack rafters.

   NOTE: A convenient way to cut the jacks in pairs is to make the cheek cut in the center of the length of the stock and the tail cuts on the ends. If a portable electric saw is to be used, it is advisable to set the saw to the plumb and cheek cut first and to make all of these cuts before the tail cuts.

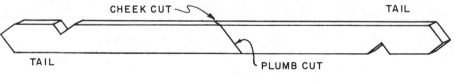

FIG. 16-15   LAYOUT OF JACK RAFTERS IN PAIRS

How to Make the Ridgeboard

1. Subtract the width of the house from the length. Using Fig. 16-16 as an illustration, the length would be 20'.

2. Add the thickness of the ridge. This would give a ridge length of 20'-3/4".

3. Starting at one end of the ridgeboard, mark every 24" for the common rafters.

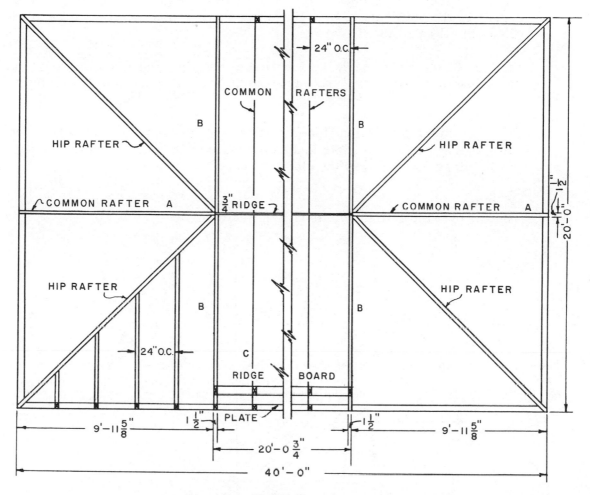

FIG. 16−16 LOCATION OF RAFTERS OF HIP ROOF

How to Place the Common Rafters

1. Erect and brace the common rafters as explained in Unit 15.

2. Erect the common rafters at the ends of the building. Nail these rafters to the plate at the seat cut and midway between the sides of the building.

3. Nail the tops of these rafters so they butt against the end of the ridge. The centerline of the edge of this rafter should be in line with the centerline of the ridgeboard.

How to Place Hip and Valley Rafters at the Plate

1. Mark the location of the hip or valley rafter at the internal or external corner of the plate as shown. These lines will cross the plate at an angle of 45° in all equal-pitch roofs.

   NOTE: These lines mark the location of the hip or valley rafters, so the centerline of the rafter will meet the outside or inside corner of the plate, A.

2. Push the hip rafter against the outside corner of the plate so the heel cut rests tightly against the plate and so the sides of the rafter are in line with the 45° lines made across the plate.

3. Toenail the rafter to the plate with three 10d nails.

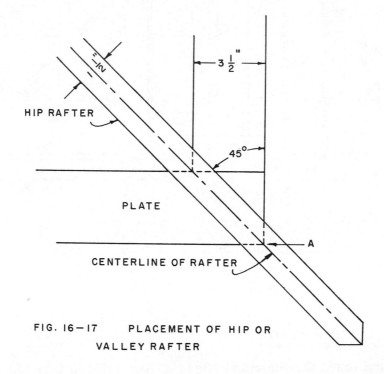

FIG. 16—17   PLACEMENT OF HIP OR VALLEY RAFTER

NOTE: If the seat cut of the hip or valley rafter is cut through to the top edge of the rafter, a centerline should be drawn along the seat cut, and the measuring line should be square across this cut as shown. The point where the centerline and the measuring line meet marks the point that is to be placed at the outside corner of the plate.

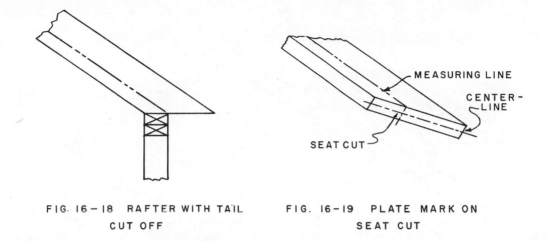

FIG. 16-18 RAFTER WITH TAIL   FIG. 16-19 PLATE MARK ON
           CUT OFF   SEAT CUT

How to Place the Hip or Valley Rafters at the Ridge

1. Nail the hip rafter into the angle formed by the two common rafters, keeping the centerline on the top edge of the hip rafter even with the top of the ridgeboard. The other three hip rafters of the roof should be placed in the same manner.

    NOTE: Valley rafters are placed so that the two outside top edges are even with the top of the two ridgeboards.

How to Place the Hip or Valley Jack Rafters

1. Make the spacing for the hip jack rafters on the plate at both sides of the four corners of the building.

2. Stretch a string, in the center of the hip rafter, from the top of the hip rafter to the bottom of the hip rafter.

    NOTE: The string should "float" above the hip rafter. This may be done by placing small blocks under the string or by tieing the string above the top of the hip rafter.

3. Nail the hip jack rafters to the plate marks at the seat cut.

4. Nail the cheek cuts to the hip rafter. If the hip rafter is not in the center of the string, pull the hip jack rafter until the hip rafter aligns with the string; then nail the cheek cut to the hip rafter.

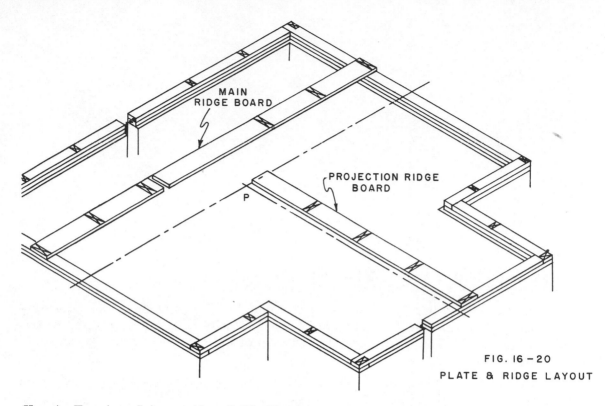

MAIN
RIDGE BOARD

PROJECTION RIDGE
BOARD

P

FIG. 16-20
PLATE & RIDGE LAYOUT

How to Erect an Intersecting Gable Roof

NOTE: Fig. 16-20 shows a floor plan of an intersecting roof with the plates marked for the common rafter locations.

1. Find the length of the main ridgeboard by measuring from the outside of the plate at one end of the building to that of the other end.

2. Cut the ridgeboard and space the plate and ridgeboard for the common rafters as explained.

3. Find the length of the ridgeboard for the projection of the building by measuring from the outside of the plate line of the projection to the centerline of the main building as shown at P. Deduct one-half the thickness of the main ridgeboard from this length.

4. Cut this ridgeboard to length.

5. Mark the spacing for the common and jack rafters on this ridgeboard, and for the common rafters on the plate.

6. Erect the common rafters and the ridgeboard as explained in Unit 15 and as shown in this unit. Be sure to brace the ridges to hold them in a level and plumb position.

7. Erect the valley rafters as explained in this unit and shown in Fig. 16-21.

8. Erect the valley jack rafters as explained in this unit and as shown in Fig. 16-22. They should be placed in such a position that the top edge of each jack, if continued, would meet the centerline of the valley rafter.

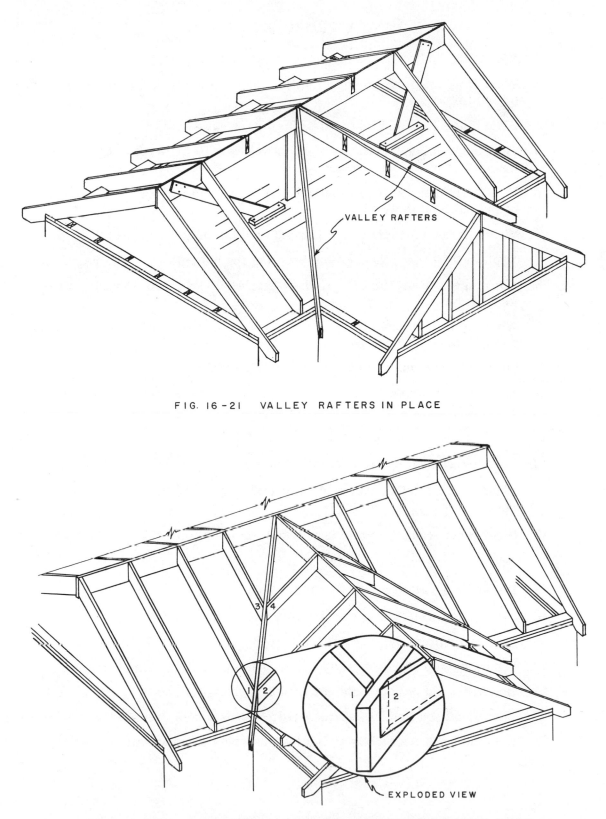

FIG. 16-21   VALLEY RAFTERS IN PLACE

FIG. 16-22   ASSEMBLY OF VALLEY JACKS (NOTE EXPLODED VIEW)

REVIEW PROBLEMS   Unit 16

1. Why is the unit of run of a hip or valley rafter greater than a common rafter?

2. What information can be gained by drawing a roof plan to scale on paper or subfloor?

3. Where are hip jack and valley jack rafters located?

4. Give the exact unit of run for a hip or valley rafter. What is the approximate number used for this unit of run?

5. How many steps will be required to lay out a hip or valley rafter if the span of the building is 22'-0"?

6. Where is the measuring line located on a hip or valley rafter?

7. Explain the two methods of intersecting the hip rafters at the ridge.

8. Describe one method for finding the length of a hip or valley rafter per foot of run.

9. When laying out the cheek or bevel cut, should the line be placed along the body or tongue of the square?

10. Why is it necessary to "back off" a hip rafter?

11. What figures would be used and which edge of the square marked to find the amount to be "backed off" a hip rafter?

12. What is meant by "dropping" a hip rafter?

13. What is a cheek or face cut on a hip rafter?

14. What other kind of rafter do the tail and seat cuts of a hip jack rafter resemble?

15. What other kind of rafter is marked for the plumb cut like the jack rafter.

16. Are jack rafters laid out in pairs?

17. Describe a valley rafter as to location.

18. State a convenient way to cut jack rafters in pairs.

19. How are rafters fastened at the plate?

20. Should the top edges of valley rafters be even with the top of the ridgeboard?

21. How can a curved hip rafter be held in a straight line?

22. How would you find the length of ridgeboards for an intersecting gable roof?

23. List the tools necessary to lay out and erect hip and valley rafters.

Unit 17   DORMERS

A dormer is a framed structure projecting from a roof surface. It might be considered a minor roof in comparison with the major roof span. It may contain a window to admit light and to permit ventilation. In some cases, it may be constructed to improve the exterior appearance of the building and to provide additional space in the interior. It may be built up from the level of the main roof plate or from a point above the plate. The front wall of the dormer may be placed back of the main building line, may project beyond it, or may be flush with it.

In many cases the dormer roof surfaces are of the same general shape as those of the main roof. The dormer does not necessarily have to carry out the main roof lines unless it is built for appearance only. Dormers constructed for the purpose of providing additional space within the building are generally of the style that will provide the most headroom. In cases where the style of dormer does not provide sufficient headroom, the plate line of the dormer may be raised above that of the main roof so that additional room is provided.

There are many styles and shapes of roof dormers. The basic type, which has the plate line raised above that of the main roof, and whose span is in proportion to the rise and run of the main common rafter, will be explained in detail. A study of this type should give the learner a general understanding of dormer layout.

FIG. 17—1   SHED DORMER

THE SHED DORMER

When framing the roof for a dormer, an opening must be left and the dormer built into this opening. If shingles are to be used on this type of dormer, the rafters must have a rise of at least six inches to the foot to permit the roof to shed rain and snow properly.

The rafters are laid out and spaced the same as a common rafter of the main roof, using the rise from the top of the dormer plate to the top of the header. The run of the rafter is taken from the outside of the dormer plate to the front face of the header.

If the dormer extends up to the ridge, the rise is figured from the top of the dormer plate to the top of the ridge. The run is figured from the outside of the dormer plate to the center of the ridgeboard.

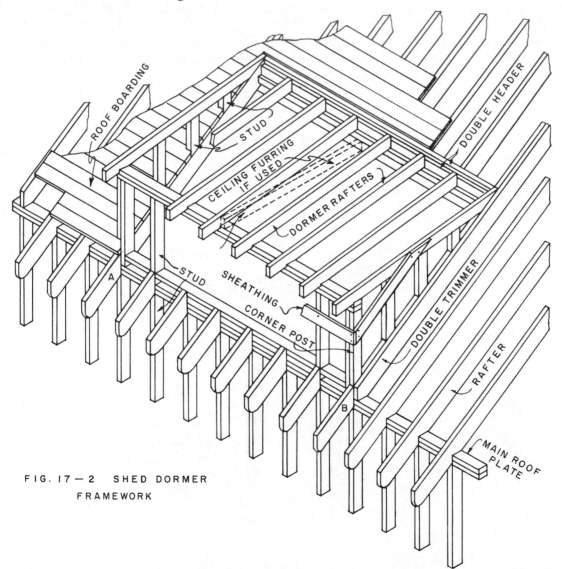

FIG. 17 — 2   SHED DORMER
FRAMEWORK

TOOLS AND EQUIPMENT FOR FRAMING DORMERS

Hammer	Spirit level	Steel square	Portable electric saw
Ripsaw	Steel tape	Sliding T bevel	
Crosscut saw	Rule	Hand axe	

How to Erect a Shed Dormer

1. Check the position of the common rafters at the sides of the dormer, see Fig. 17-2 A and B.

2. Build the front wall section of the dormer. The studs, plates, and the window opening are framed and erected as in an outside wall of the main building.

3. Raise the wall and toenail the studs to the main plate. Brace and plumb the wall both ways as described in Unit 10. Apply the sheathing to the front wall.

4. Install the header for the dormer rafters at the point where they will meet the main roof.
   NOTE: Sometimes the header is omitted and the dormer rafters are extended to the ridgeboard of the main roof. In such a case, the ridgeboard should be reinforced to carry the load of the shed roof rafters.

5. Lay out and cut the required number of dormer rafters as explained.

6. Space and spike the rafters in place.

7. Double the header in the same manner as the header is doubled in floor joists.

8. Double the main common rafters on each side of the dormer.
   NOTE: If the side wall studs of the dormer do not rest on the common rafters but are supported from the joists below, it is not necessary to double the rafters.

9. Cut the tail rafters needed to carry the cornice across the lower front of the dormer.

10. Space and spike the rafter tails securely to the front wall of the dormer. Use a straightedge to keep the rafter tails in line with the main common rafters.

11. Lay out the spacing for the side studs 16 inches o.c.

12. Hold a piece of 2×4 vertically at a spacing mark. Mark the top and bottom cuts on the 2×4, using the underside of the dormer rafter and the top side of the main common rafter as guides for the pencil.
    NOTE: This is similar to marking gable studs as shown in Unit 15.

13. Repeat this operation for the other side wall studs.

14. Cut the studs and nail them in place. Apply sheathing to sides of the dormer.

THE GABLE DORMER

The gable dormer sheds the rain and snow two ways. This type is often used in preference to a shed dormer when it is necessary to provide more pitch for the dormer or to carry out the general lines of the main roof. A hip end could be used instead of the gable.

The ridge of this dormer is lower than that of the main roof. In this case a header is installed between the rafters of the main roof at the point where the ridge-board and the valley rafters of the dormer meet the main roof (C).

The rafters are laid out in the same manner as the common, valley, and valley jack rafters of the main roof except that the rise and span of the dormer are used.

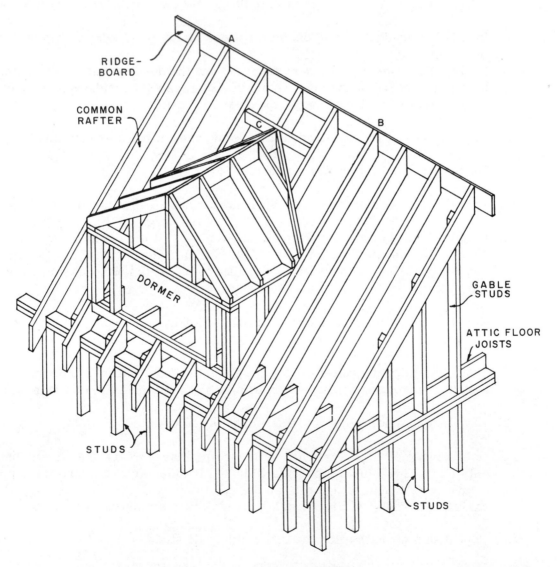

FIG. 17-3   GABLE DORMER

To avoid having a pitch on the dormer roof different from that on the main roof, the same rise and run per foot should be used on the dormer common rafter as are used on the main common rafter.

If the dormer plate is to be raised above the main plate, it should be placed in accordance with the steps marked on the pattern of the main common rafter.

Assume that the run of the main roof common rafter is 14′ and the rise per foot of run is 9″. To lay out the common rafter, 14 steps of 9″ on 12″ would be taken,

If the plates of the dormer are to be 6′ or 72″ above the main plate, the top of the dormer plate will come at the end of the 8th step taken with the square along the main common rafter. (72″ ÷ 9″ = 8 steps.) The line showing the level cut of the 9th step will show the top of the double plate of the dormer.

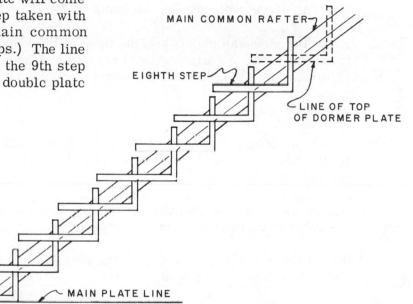

FIG. 17-3   HEIGHT OF DORMER PLATE

If the dormer ridge is level with the main ridge, the length of the common rafter of the dormer can be found by subtracting these 8 steps from the total of 14 steps on the main common rafter. This will leave 6 steps of 9″ on 12″. Thus the run of the dormer rafters will be 6′, and the total width of the dormer will be 12′. The half thickness of the ridgeboard should be deducted the same as for the main common rafter.

In many cases, the dormer ridge is below the main ridge. As in the previous example, 8 steps should be deducted from the length of the main common rafter if the dormer plate is to be 6′ above the main plate. In addition, a number of steps corresponding to the difference in height between the two ridges should also be deducted. Assuming that this difference is 18″, and since the rise per foot of run is 9″, this additional deduction would be 2 steps. The run of the dormer common rafter would then be 4′ and the width of the dormer between the rafters would be 8′.

After finding the length of the dormer common rafter, the lengths and cuts of the remaining rafters of the dormer may be found in the manner explained for similar rafters of the main roof.

How to Erect a Gable Dormer Using a Header

1. Check the width of the dormer to see that the common rafters, A and B, are properly placed.

2. Erect the front wall section the same as for the shed dormer.

3. Erect the side wall plates, keeping them at the same height as the top of the front wall plate, and allowing them to extend to the sides of the main common rafters. Spike the plates to the sides of these common rafters.

4. Erect the side wall studs.

5. Erect the common rafters of the gable dormer in the same manner as the main common rafters. Allow the ridgeboard to extend beyond the face of the outside plate.

6. Plumb the outside gable rafters with the face of the outside wall section of the gable and brace them in this position.

7. Erect the studs from the top of the outside wall plate to the underside of the gable common rafters.

8. Level the top of the dormer ridgeboard.

    NOTE: If the side wall plates are level and plumb and the common rafters of the dormer have been properly erected, the ridgeboard should be level.

9. Install the valley rafters, following the same general procedure as in erecting a full-length valley.

10. Cut the ridgeboard off at the inside ends of the valley rafters. This point gives the location of the header.

11. Install the header, keeping the top edge of the header flush with the top of the main common rafters.

12. Install the valley jack rafters of the dormer and the main roof.

13. Install the short common rafter from the header to the main ridge.

14. Sheath the front and side walls of the dormer.

FRAMING A GABLE DORMER WITH A LONG AND SHORT VALLEY RAFTER

The upper end of a small gable dormer may be framed as shown previously, while a large gable dormer may be framed into the main roof by using a long and a short valley rafter as shown below. This roof plan shows a main roof with a common rafter having a run of 12′ and a rise of 9″ per foot of run. The gable front wall is flush with the main building line. The run of the gable common rafter is 10′.

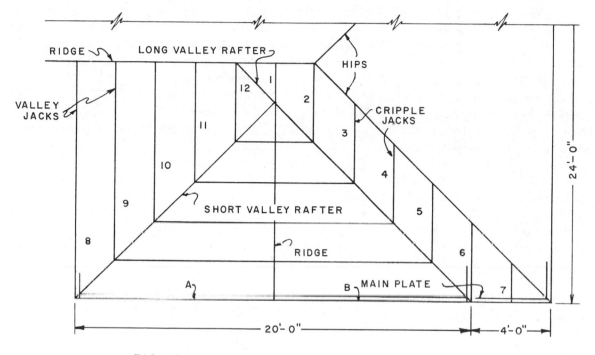

FIG. 17 — 4   LONG AND SHORT VALLEY RAFTERS

A. Length of Long and Short Valley Rafters

The long valley rafter is laid out in the same general way as a valley rafter of a main roof of 24′ span (12 steps of 9″ and 17″). A single cheek cut is used at the ridge as shown.

The short valley rafter is laid out in a similar manner except that only 10 steps are taken with the square because 10 steps were taken in laying out the gable common rafter.

B. Cuts of Long and Short Valley Rafters

The top plumb cut of the short valley rafter is found by using the rise and run of the valley rafter (9 on 17). Mark on 9 for the plumb cut. The cheek cut should be square, as this rafter butts against the long valley rafter. Half the thickness of the long valley rafter is deducted from the length of the short valley rafter in the same way as half the thickness of the main ridgeboard is deducted from the length of the long valley rafter.

C. Layout of Ridgeboard

The length of the ridgeboard is found by plumbing down from the intersection of the long and short valley rafters and measuring a level distance to the outside of the main wall plate. The level distance gives the length of the ridgeboard.

D. Lengths and Cuts of Jack Rafters

Since rafter 1 is parallel to the main roof common rafters, and since the jack rafters are spaced 24″ o.c., the length of this rafter is equal to 2 steps on the common rafter of this roof. The top or ridge cut is the same as that of the common rafter. The bottom or valley cut is a combination of a plumb cut, which is the same as at the top of the rafter, and a cheek cut. This cheek cut may be found by measuring the diagonal of the rise and run of the common rafter on the square. The rise in this case is 9″ and the run is 12″. The diagonal will be found to be 15″. This figure is then taken on the tongue of the square, and 17″ is taken on the body. A mark made along the outside of the tongue will show the cheek cut.

The cripple jacks marked 2, 3, 4, 5, and 6 are laid out in the same way except that the combination plumb-and-cheek cut is used at both ends. The faces of these two cuts on each cripple jack must be parallel to each other. The length of these rafters is equal to 4 steps on the main common rafter. This length is used because the corner of the gable is 4′ from the end of the building, and the run of the jacks between the valley and hip is 4′.

The hip jack marked 7 is laid out as described in Unit 16. The cheek cut is the same as for the other jacks, and the seat and tail cuts are the same as the common rafter cuts.

The valley jacks (8, 9, 10 and 11,) are laid out in the same manner as valley jacks of any main roof. The length of rafter #8 is 12 steps of the square, rafter #9 is 10 steps, rafter #10 is 8 steps, and rafter #11 is 6 steps.

Rafter #12 is 4 steps in length. The top and bottom cuts of this rafter consist of a plumb cut and a cheek cut similar to those of rafter #2 except that the cheek cuts converge toward one another.

How to Erect a Gable Dormer Using a Long and a Short Valley Rafter

NOTE: The long and short valley rafters extend to the main plate of the building, so there are no gable side walls. The front wall would be studded and sheathed the same as any gable.

1. Lay out and cut the long and short valley rafters as described.

2. Lay out and cut the valley jack rafters for both the main and gable roofs.

3. Install the long valley rafter first. The procedure is the same as for the main roof valley rafter. See Unit 16.

4. Install the short valley rafter, making sure that the seat cut is located on the plate so that the measuring line of the valley rafter intersects the outside edge of the plate.

5. Nail the dormer ridgeboard temporarily at the intersection of the long and short valley rafters.

6. Nail the two common rafters (A and B) to the main plate in the proper location.

7. Plumb up from the outside face of the main plate to the ridgeboard and mark it at this point. This locates the outside face of the common rafter at the ridge.

8. Cut the ridgeboard along this plumb line and nail the common rafters to the ridge.

9. Check the top of the ridgeboard from the outside gable end back to the point where it intersects the long and short valley rafters to see if it is level. When it is level nail the ridgeboard firmly to the valley rafters.

10. Install the remaining gable and main roof valley jack rafters as described in previous units.

How to Install Cripple Jack Rafters

1. Lay out and cut the cripple jack rafters as described.

2. Space these rafters by lining them up with the valley jack rafters of the gable on the opposite side of the valley.

NOTE: If the hip rafter has been dropped, place the tops of the cripple jacks even with the top of the hip rafter. If the hip has been backed, place the cripples against the hip rafter as shown in Unit 16.

OPENINGS IN ROOF SURFACES

Chimney holes, skylights, and scuttles are framed in roof rafters in a similar manner to openings in floor joists. The headers between the rafters are placed so that their faces are plumb. The opening in the roof rafters is located by plumbing up from the face of the headers and trimmers of the opening in the floor joists and by placing headers between the rafters as shown. (Fig. 17-5.)

When a chimney hole, a scuttle, or a skylight is to be located, a plan of the opening is often drawn full size on the attic floor. The points showing the inside dimensions of the opening are then plumbed up to the rafters or to the boards nailed temporarily to the roof.

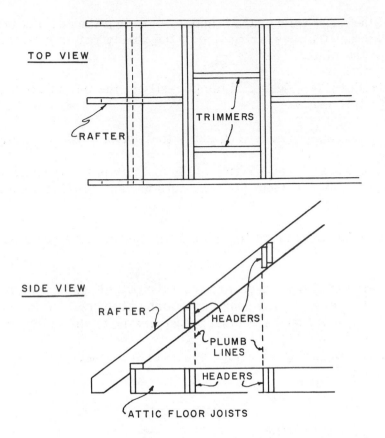

FIG. 17-5   CHIMNEY OPENING IN ROOF

How to Provide for a Chimney Opening

NOTE: It is assumed that the opening in the attic floor joists is framed to the size of the chimney.

1. Plumb a line from the corners of the opening in the attic floor joists to the rafters with a spirit level and straightedge or with a plumb bob.

2. Mark a plumb line on the rafters to be cut.

3. Nail a strip across the tops of the rafters to be cut and across at least two rafters on each side. This will support the rafters after they have been cut through.

4. Cut the rafters along the plumb line.

5. Spike the single headers to the rafters.

6. Spike the double headers in place against the single header. Note the arrangement at the tops of the rafters.

7. Plumb up from the trimmers in the attic floor and mark their position on both the double headers in the roof opening.

8. Lay out and cut the trimmers. The cut on each end of a trimmer is the same as the ridge cut of the common rafter.

9. Spike the trimmers in place, keeping the tops of the trimmer on a line with the tops of the common rafters.

REVIEW PROBLEMS   Unit 17

1. What is a dormer, and what purposes might it serve?

2. Do dormers have to carry out the shape of the main roof?

3. What is the minimum pitch suggested for a dormer roof if asphalt shingles are to be applied?

4. When the rafters of a shed dormer extend to the ridge, what precautions should be taken?

5. When is it necessary to double the main common rafters on each side of a dormer?

6. What is the difference between a gable dormer and a shed dormer?

7. How can you be sure the pitch of the roof of a dormer will be the same as that of the main roof?

8. What is meant by framing a gable dormer with a long and a short valley rafter?

9. Where are cripple jack rafters located?

10. How are the openings for dormers located in a sloped roof?

11. List the tools necessary to frame and erect dormers.

Unit 18   ROOFS OF UNEQUAL PITCH

An unequal-pitch roof is one in which the main part has a different pitch from that of a projecting part.

If the rise per foot of run of a common rafter of a projection is not the same as that of the common rafter of the main roof, the two roofs will have unequal pitches. This happens under the following conditions:

1. When the ridges of the two parts of the roof are the same height and the spans of the two are different.

2. When the ridges are unequal in height (the rises different) and the two spans are the same.

3. In some cases, when ridges are unequal in height and spans are different.

In the preceding units, the only roofs considered (with the exception of the shed dormer) were those whose main and intersecting parts had the same slope.   For example, the gable dormer had the same rise per foot of run as the main roof.

The width or height of an intersecting roof of the hip or gable type is sometimes restricted. In this case, the rise per foot of run of the main common rafter may be different from that of the gable common rafter. A roof of this type presents more difficult layout problems than the roof whose parts all have the same slope.

LAYOUT OF A ROOF OF UNEQUAL PITCH

Assume that a building is to be 24′ wide and 40′ long, with a roof projection on the side (Fig. 18-1). The run of the common rafter of the main roof will be 12′, and that of the projection common rafter will be 6′. The rise of each of the common rafters is 8′, which brings the ridges of the intersecting roof and of the main roof even.   The rafters are to have no overhanging tails for a cornice. It is often advisable to draw a scaled plan of the rafters showing their location in the roof.

A. Layout of Common Rafters of the Projecting Roof

The rise per foot of run of the common rafter is found by dividing the total rise of the rafter by the run of the rafter. (96″ ÷ 6′ = 16″ rise per foot of run.) Since 16″ is too far out on the tongue of the square to permit convenient holding, the 16″ rise is taken on the body of the square and the 12″ run is taken on the tongue.

If the measuring line were placed below the top of the rafter as in the equal-pitch roof, the tops of the plumb cuts of the main and the projection common rafters at the plate line would not be the same height. The top of the projection rafter would be higher than that of the main roof rafter. Therefore, the measuring line for rafters in unequal-pitch roofs is at the top edges of the rafters if there are to be no rafter tails.

154

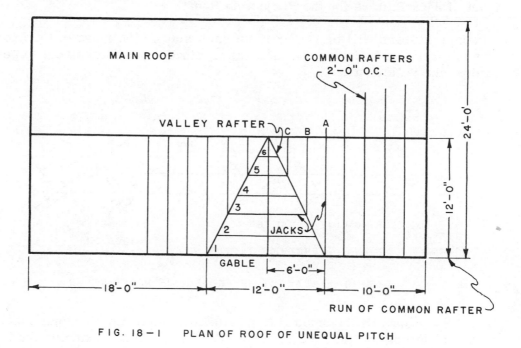

FIG. 18—1   PLAN OF ROOF OF UNEQUAL PITCH

The length of the projection common rafter is found by stepping the square along the top edge of the rafter with 16 on the body and 12 on the tongue. Six steps should be taken. If the rise per foot of run comes out in a fraction, such as 18 2/3", it can be measured accurately by using the back of the body of the square where the inches are divided into twelfths.

The ridge or plumb cut is marked along the outside of the body, and the seat or level cut along the outside of the tongue. The deduction for half the thickness of the ridgeboard is made in the same manner described in Unit 15.

When the span of the main or projection roof is in odd feet, the method of finding the common rafter length is the same as described in Unit 15.

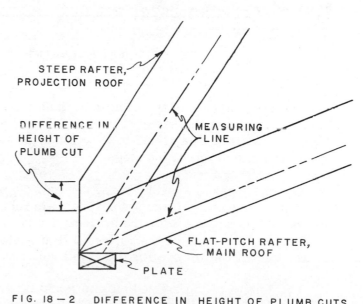

FIG. 18—2   DIFFERENCE IN HEIGHT OF PLUMB CUTS WHEN USING MEASURING LINE

B. Layout of Jack Rafters for the Projection Roof

Six jack rafters will be required on each side of the ridge of the projection as shown. Rafter 1 is considered a jack rafter, since it has a cheek cut to permit it to fit against the valley rafter.

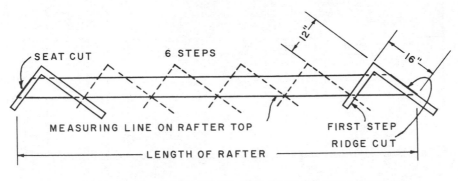

FIG. 18-3   LAYOUT OF PROJECTION ROOF RAFTER

Fig. 18-4 shows the common rafter pattern divided into six equal spaces so it can be used in laying out the jack rafters. For the purpose of finding the length of the rafter, the top edge of the rafter is used.

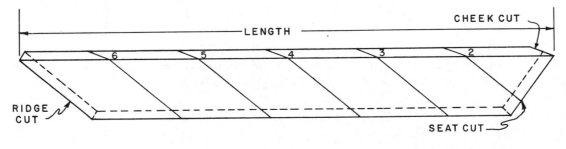

FIG. 18-4   LENGTHS OF JACK RAFTERS

The longest jack is the same length as the common rafter pattern. The ridge cut and seat cut are the same as those of the common rafter. A cheek cut will have to be made to allow this rafter to fit against the valley rafter. The plumb cut is the same as the plumb cut at the ridge.

The cheek cut is found by taking one-half the span of the projection on the tongue of the square and one-half the span of the main roof on the body of the square. In this case, these figures are 6 for the projection and 12 for the main roof. The square is placed on the top edge of the rafter in this position, and a line is marked along the outside edge of the tongue for the cheek cut.

The length of the second jack is found by measuring from the top of the ridge cut of the first jack along the edge of the rafter to line 2.

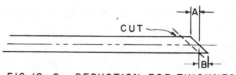

FIG. 18—5 CHEEK CUT FOR JACKS

The length of the third jack is measured from the ridge cut to the line marked 3. The remaining jacks are measured the same way.

The deduction for half the thickness of the valley rafter at the valley end of the jack is found by gaging a centerline along the top

edge of the jack and squaring a line across this centerline at the point where it meets the cheek cut. The distance from this line to the short or long tip of the cheek cut (see A or B) is the deduction to be made. This deduction is measured along the centerline and back from the face or cheek cut. See dotted line.

FIG. 18—6   DEDUCTION FOR THICKNESS OF VALLEY RAFTER

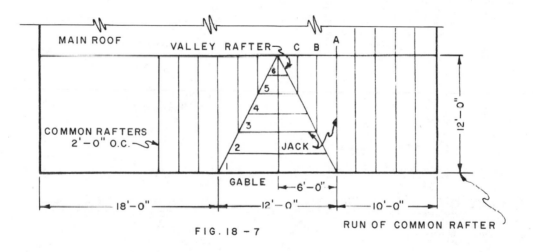

FIG. 18 — 7

C. Layout of Rafters of Main Roof

The length and cuts of the main roof common rafters are found in the same manner as those of the projection roof common rafters.

The length of the main roof jacks is found in the same way as those of the projection roof. In Fig. 18-7 the valley jack rafters needed for the main roof at one side of the projection are marked A, B, and C. The jack rafter A is the full length of the common rafter of the main roof. The lower end fits against the valley rafter in the same way as the corresponding jack of the projection.

The lengths of the other two main roof jacks are found by dividing the length of the main common rafter into three equal parts. This is because there are only three valley jacks required for this section of the roof instead of six as in the case of the projection roof jacks.

The same general procedure is followed in finding the cuts of the main roof valley jacks as for the projection roof valley jacks. The only difference is that the cheek cut for a main roof jack is marked along the body of the square instead of along the tongue.

D. Layout of Valley Rafters

The length of a valley rafter of an unequal-pitch roof is found in the same way as that of any valley rafter except that the unit of run for a valley rafter of an unequal-pitch roof is figured in terms of the two unequal pitches of the roof. The unit of run for a valley rafter in a roof of equal pitch is 17″. The unit of run of the valley rafter for an unequal-pitch roof is the diagonal of one-half the span of the main roof and one-half the span of the projection roof. In Fig. 18-9, one-half the projection roof span is 6′ and one-half the main roof span is 12′. Therefore, the diagonal is measured from 6 on the tongue of the square to 12 on the body and will be found to be 13 7/16″. The square is stepped along the top edge of the rafter 12 times, using a rise of 8″ on the tongue of the square and 13 7/16″, the unit of run, on the body. The ridge plumb cut of the valley is marked along the outside edge of the tongue, and the seat cut is marked along the outside edge of the body.

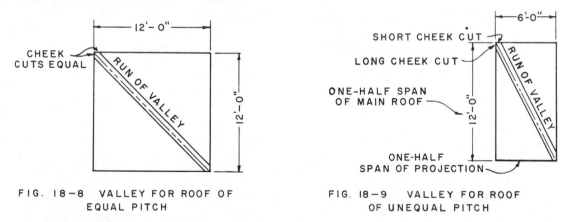

FIG. 18-8   VALLEY FOR ROOF OF   FIG. 18-9   VALLEY FOR ROOF
EQUAL PITCH   OF UNEQUAL PITCH

The method of finding the cheek cuts at the ridge for the equal-pitch valley rafter is described in Unit 16. The cheek cuts on this valley rafter are of equal length because the run of the valley is the diagonal of the main roof and projection roof spans, which are equal (Fig. 18-8).

The two parts of a roof of unequal pitch have unequal spans. Therefore, the run of the valley rafter is the diagonal of a rectangle instead of a square. The cheek cut at the ridge on one side of the rafter will be different in length from the cheek cut on the other side (Fig. 18-9).

One of the cheek cuts at the ridge end of the valley rafter is the same as the cheek cut of the main roof valley jack. The other cheek cut is the same as the cheek cut of the projection roof valley jack. In the roof illustrated in Fig. 18-9 the long cheek cut will come against the projection roof ridge and the short cheek cut against the main roof ridge. The same cuts are used at the lower end of the valley rafter at the plate. However, the positions of the two cuts at the plate end are reversed from the position of the two cuts at the ridge end (Fig. 18-9).

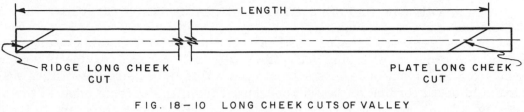

FIG. 18—10   LONG CHEEK CUTS OF VALLEY
RAFTER

These cuts at the two ends of the valley rafter can be more easily laid out by using the jack rafters as patterns than by using a square. The reason for this is that the intersection of the two cheek cuts on the valley rafter is not on the centerline, as one cut is longer than the other. The cheek cut of the main roof jack rafter is used as a pattern for marking the long cheek cut on the valley rafter. After the length of the valley rafter has been stepped off, this long cheek cut should be marked (Fig. 18-10). The cheek cut of the projection roof jack rafter is used for marking the short cheek cut on the valley rafter. This short cheek cut is marked on both ends of the valley rafter (Fig. 18-11).

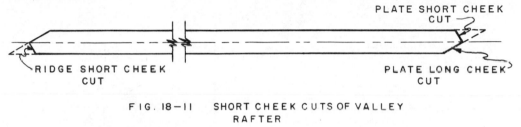

FIG. 18—11   SHORT CHEEK CUTS OF VALLEY
RAFTER

The deduction from the length of this valley rafter for the thickness of the ridgeboard is shown in Fig. 18-12. An amount equal to one-half the thickness of each ridgeboard is taken off the corresponding cheek cut.

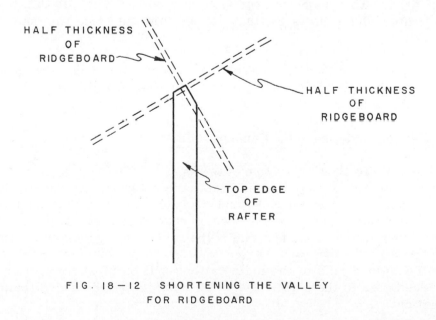

FIG. 18—12   SHORTENING THE VALLEY
FOR RIDGEBOARD

159

LAYOUT OF RAFTERS WITH OVERHANG FOR CORNICE

When it is necessary to have a continuous cornice on both sections of the unequal-pitch roof, the problem of laying out rafters with tails for the cornice is somewhat different from the layout of rafters without tails. This is because the rafters with no tails are laid out according to the span measured from the outside edge of the plate (see A, Fig. 18-13), whereas the rafters with tails are laid out according to the span measured from the outside vertical end of the rafter tail (see A, Fig. 18-14). The rafters with tails are laid out from this point to provide a level point common to both pitches of the roof.

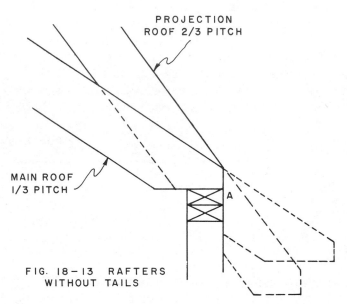

PROJECTION
ROOF 2/3 PITCH

MAIN ROOF
1/3 PITCH

A

FIG. 18—13 RAFTERS
WITHOUT TAILS

For example, assume that the main roof common rafter has a pitch of 8″ on 12″ or 1/3 pitch. The projection common rafter has a pitch of 16″ on 12″ or 2/3 pitch. If the rafters have no tails, they can be laid out from the outer edge of the plate as shown.

If they are to have tails, it will be impossible to run a continuous cornice from the 1/3-pitch rafter to the 2/3-pitch rafter because the tops of the rafters are not level with one another at the ends of the rafter tails.

A. Added Run and Rise of the Common Rafters

Fig. 18-14 shows a section of the same roof, but provision is made for an overhanging cornice. If the cornice is to have a 12″ run as shown, the rafter tail will project one step of the square beyond and below the plate line. It will be necessary to add one step of the square to the common rafter when it is stepped off. This is sometimes referred to as the added run of the rafter to provide for the run of the cornice.

The added rise of the main rafters will be 8″ because both ridges are even. The added step to the main rafter will bring the rafter tail 8″ below the main plate or seat cut.

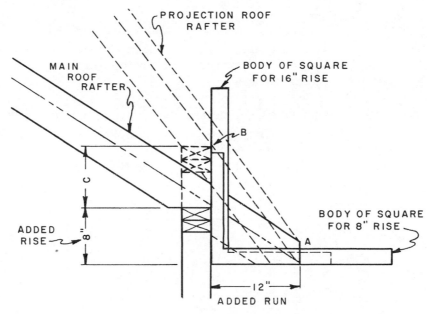

FIG. 18—14   ADDED STEP OF SQUARE ON RAFTERS

The tail of the projection roof rafter, which has a rise of 16″ per foot of run, will also be one step below the seat cut. The plate for this rafter should be raised 8″ above the main roof plate so both rafter tails will be on the same level.

B. Layout of the Main Roof Rafter

The measuring line of this rafter may be located in the same way as described for the common rafter of the equal-pitch roof. The figures on the square will be 8″ rise and 12″ run. The method of finding the length, cuts, and allowance for the ridge-board is the same as described in Unit 15.

C. Layout of Projection Roof Rafter

The measuring line of the projection roof rafter is found by placing the square on the rafter, using a rise of 16″ on the body and a run of 12″ on the tongue, and marking a plumb cut at the end of the rafter stock (A, Fig. 18-14). The distance from the top of the rafter down along this plumb line must be the same as the corresponding distance on the main roof rafter. This point establishes the measuring line of the projection roof rafter.

The remaining layout of this rafter is similar to the layout of the main roof rafter. The seat cut is marked out on the projection roof rafter at the first step of the square, where the outside edge of the body of the square crosses the measuring line of the rafter (B, Fig. 18-14).

The distance the plate of the projection is raised over that of the main roof is shown at C. This distance is the difference between the rise per foot of run of the two rafters, 8″ in this case.

161

D. Layout of Valley Rafter

The run of the valley rafter is also increased by the length of the overhanging cornice, but this added distance is measured along the diagonal of the two unequal spans of the main and projection roofs.

A simple method of showing the location and diagonal length of this rafter is to draw a plan view of the roof, showing the plate, ridge, and cornice lines at the valley intersection. The drawing may be made to a convenient scale, Fig. 18-15, or chalk lines may be snapped on the attic subfloor of the building showing a full-size plan of the rafters.

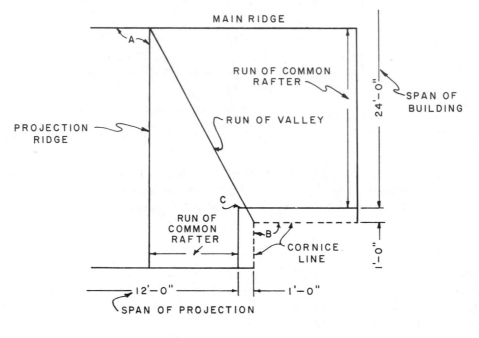

FIG. 18 — 15   RUN OF VALLEY RAFTER

Notice that the run of the valley rafter is measured along the diagonal line that extends from the intersection of the two ridgeboards (A) to the intersection of the two cornice lines (B). The valley rafter does not pass over the intersection of the two plate lines (C), because the runs of the two common rafters are unequal. But the runs of the cornices are equal on both the projection and the main roofs.

The method of locating the measuring line on the valley rafter is the same as that described for the projection common rafter in this unit, except that the rise and run of the valley rafter are used instead of the rise and run of the projection roof rafter. If a measuring line is used below the top edge of the projection and main common rafters, the measuring line on the valley rafters should be located in the same way.

The length of the valley rafter is found in the same manner as the length of the unequal valley rafter with no tail, except that the added run on the two common rafters forms a rectangle, from which the increased run of the valley rafter is found.

Since the run of the projection roof rafter is 6', the run of this rafter, including the overhanging tail of 1', will be 7'. The run of the main common rafter with overhang will be 13'. The length of the valley rafter will be the diagonal of a rectangle 7' by 13'. The remaining problem of finding the length of the rafter is similar to finding the length of the valley rafter with no tail.

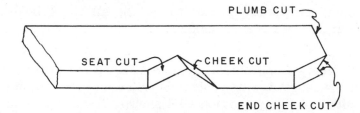

FIG. 18—16 VALLEY RAFTER TAIL AND SEAT CUTS

The cheek cuts at the ridge are found as in the case of the rafter without a tail. The seat cut is found as described for the main common rafter with a tail. The plumb cut, which fits against the plate at the seat of the rafter, also has a cheek cut (Fig. 18-16). When the valley rafter lies on the main roof side of the corner of the plate line, this cheek cut is parallel to the ridge cheek cut at the main ridgeboard. When the rafter lies on the projection roof side, the cheek cut is parallel to the projection ridge cheek cut. The cheek cuts at the end of the rafter tail are found in the same way as those at the ridge. However, the left cheek cut at the top of the valley rafter is used for the right cheek cut at the lower end and vice versa.

TOOLS AND EQUIPMENT

Ripsaw   Spirit level
Crosscut saw   Framing square
Hammer   Rule or steel tape
Sliding T bevel   Chalk line

How to Erect Intersecting Roofs of Unequal Pitch

NOTE: The erection of a roof of unequal pitch differs slightly from that of the equal-pitch roof because the rafters are laid out somewhat differently. The ridgeboards, plates, common rafters, and valley jacks are laid out and spaced in the same general way as for any equal-pitch roof. The cuts of the various rafters have been described in previous units pertaining to each particular type of rafter.

1. Lay out the ridgeboards and plates of the main and projection roofs in the same general way as described in Unit 16.

2. Lay out and cut the common rafters of the main roof and projection roof in the same manner as the common rafters of the gable roof. See Unit 15.

3. Lay out and cut the valley jacks in the same manner as described in Unit 16.

   NOTE: Some carpenters prefer to use the sliding T bevel to transfer the bevel cuts from one rafter to another. This method makes for more accurate work.

   Fig. 18-17 shows a convenient way of arranging the rafters on edge in order to lay out the lengths, ridge cuts, and valley cuts and to make them in pairs for both sides of the projection roof shown.

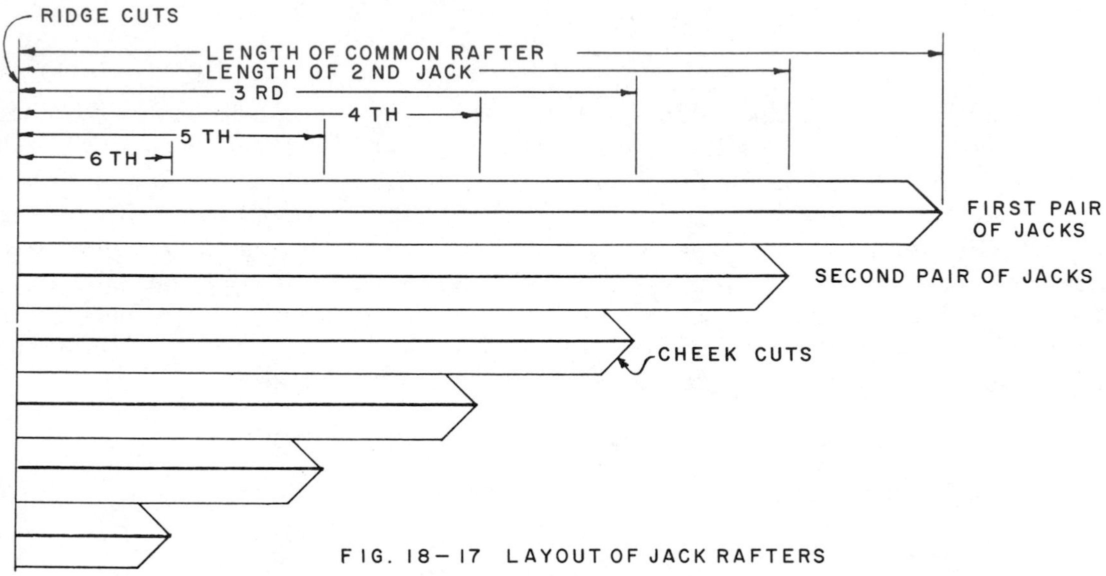

FIG. 18-17 LAYOUT OF JACK RAFTERS

4. Lay out and cut the valley rafters as described.

   NOTE: The above procedure applies to rafters with or without rafter tails.

5. Erect and brace the ridgeboards and common rafters of the main section of the roof as shown in Unit 15.

   NOTE: The roof plan, Fig. 18-7, Page 157, shows there is no need to raise the plate of the projection roof section, because the projection does not extend beyond the side wall of the building and there are no rafter tails. When the projection extends beyond the side wall and there are rafter tails, that section of the plate must be raised to form a seat support for the common rafters.

6. If necessary, raise the plate for the steeper-pitched rafters by doubling up the plate. If the rise is over 6″, build up the plate.

7. Erect and brace the ridgeboard and common rafters of the projection part of the roof.

8. Erect the valley rafters as described in Unit 16.

9. Install the valley jack rafters as shown in Unit 16.

How to Erect Intersecting Trussed Roofs

NOTE: The construction of a roof truss can be done on a jig table or directly on the floor of the house. The erection is simple because the trusses are fairly light in weight. They can be placed in position upside down and then tipped up into their proper places and fastened. The intersections of roofs on L-shaped or more complicated buildings are not as easily handled as those of roofs on simple rectangular houses. The intersecting area may be built up by using small preassembled roof trusses or by conventional construction employing small precut jack rafters. In either case, the intersecting section is placed on the top of the roof sheathing. A one-inch thick plate is laid on the sheathing of the main roof at the angle of intersection, and the trusses or individual jack rafters and ridgeboard are fastened to it.

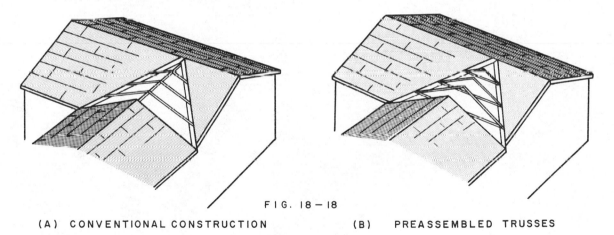

FIG. 18-18

(A) CONVENTIONAL CONSTRUCTION   (B)   PREASSEMBLED TRUSSES

REVIEW PROBLEMS   Unit 18

1. Describe a roof of unequal pitches.

2. Explain two conditions which would cause roofs to have unequal pitches.

3. Which type of roof presents the more difficult layout problem, one of equal or one of unequal pitch?

4. How would you find the rise per foot of run of the projecting roof of unequal pitch?

5. If there are to be no rafter tails on a roof of unequal pitch, where would the measuring line be placed?

6. Is the run of a valley rafter for a roof of unequal pitch the diagonal of a square or a rectangle?

7. Are the cheek cuts of valley rafters in an unequal-pitched roof the same?

8. Why is it impossible to run a continuous cornice from a 1/3-pitch rafter to a 2/3-pitch rafter?

9. Give the meaning of "added run of a rafter" when laying out a cornice for a roof of unequal pitch.

10. When is it necessary to have two different plate heights in roofs of unequal pitch?

11. Describe two methods of intersecting trussed roofs.

12. Are the intersecting members of a trussed roof placed on the top of the sheathing?

13. List the tools necessary to frame and erect roofs of unequal pitch.

Unit 19   SPECIAL FRAMING PROBLEMS

There are some special types of roof construction which have not been described in the previous units on roof framing. These deviations from the regular type of roof are often used by the designer to carry out certain lines of the roof and to modify the exterior appearance of the house. The carpenter should have a general knowledge of their layout and construction.

ODD-PITCH RAFTER

An odd-pitch rafter is one where the total rise is not an even fractional part of the total span of the building. For example, a rafter may have a total rise of 12'-8", and the span of the building may be 26'. The cuts for the common rafter would be difficult to lay out by using the usual 12" unit of run as described in the previous units on roof framing because the rise per foot of run will not come out in even inches and fractions of an inch.

The outside edge on the back of the square is divided into twelfths of an inch for the purpose of laying out rafter cuts of fractional-rise rafters. When 1" on the square represents 1' of rise or run on the rafter, 1/12" will represent 1" of rise or run.

The method employed to lay out the rafter for the building in the above example is to use the total rise of 12'-8" or 12 8/12" on the tongue of the square and 13" on the body. The 13" represents the total run of the rafter. The square is stepped along the rafter 12 times, using 12 8/12 on 13 in the same manner as for the even-pitch common rafter.

When the figures of the total rise and run are greater than the length of the square, both figures may be divided by two. The diagonal must then be doubled to find the true length per foot of run of the rafter.

EXAMPLE: Assume that the total rise of the rafter is to be 20'-4" and the total run 14'. Dividing these figures by 2 and changing the result to inches gives 10 2/12" on the tongue of the square and 7" on the body. The diagonal of these measurements (12 3/8") is doubled to find the true line length of the rafter.

The square could also be reversed and 20 4/12" used on the body of the square and 14" on the tongue, the square being stepped along the rafter twelve times.

TOOLS AND EQUIPMENT FOR SPECIAL FRAMING PROBLEMS

Hammer   Steel square
Crosscut and ripsaws   Sliding T bevel
Spirit level   Rule

How to Erect an Odd-Pitch Rafter

There should be no difference between spacing the ridgeboard and plates and cutting and erecting the rafters and similar operations for even-pitch rafters.

THE SNUB GABLE

A snub gable is made up of two shortened common rafters running into a level plate. This plate is located at some point between the main plate and the ridge. The size of the snub gable is determined by the height of the gable plate above the main plate.

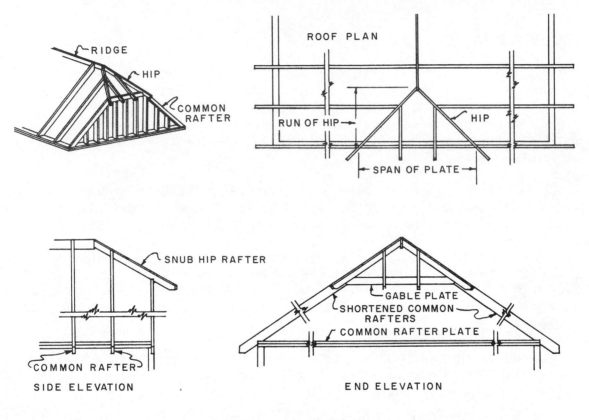

FIG. 19 — I   SNUB GABLE

The height of the gable plate above the common rafter plate is found by using the stepoff method on the common rafter. The number of steps is determined by the rise in feet of the common rafter.

EXAMPLE: Assume that a main roof is to have a span of 24′ and a rise of 8′. The top of the gable plate is to be 6′ above the common rafter plate. A common rafter would be laid out with **12** steps of **8″** on **12″**.

A. Height of Gable Plate

The number of steps of the square to be taken on the shortened common rafter from the main plate to the gable plate may be found as follows: 72″ (total rise to top of gable plate) ÷ 8″ (rise per foot of run) = 9, so 9 steps should be taken (Fig. 19-2).

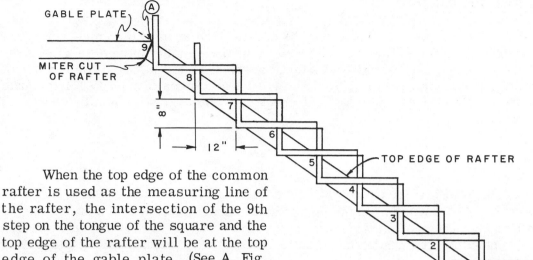

FIG. 19-2   HEIGHT OF GABLE PLATE

When the top edge of the common rafter is used as the measuring line of the rafter, the intersection of the 9th step on the tongue of the square and the top edge of the rafter will be at the top edge of the gable plate. (See A, Fig. 19-3.) When measuring line of the rafter is below the top edge, an allowance must be made for the vertical distance to the top of the common rafter.

B. Length of Gable Plate

The length of the gable plate will be twice the difference between the run of the shortened common rafter and the run of the.full common rafter. Since in this case this difference is 3 steps of the square, or 3' of run, the length of the gable plate will be 6'. This distance is shown as Span of Plate in Fig. 19-1.

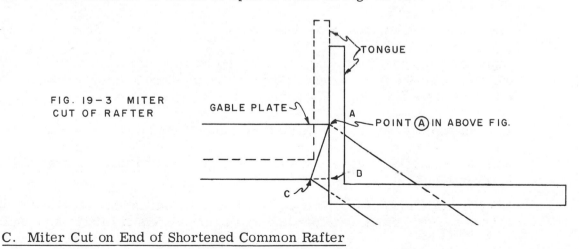

FIG. 19-3 MITER CUT OF RAFTER

C. Miter Cut on End of Shortened Common Rafter

A miter cut must be made on both the shortened common rafter and on the gable plate at the point at which they join (Fig. 19-3). This cut is found by measuring down from A, along the outside of the tongue of the square at the 9th step, a distance equal to the width of the gable plate. This will locate point B. A line should then be squared from line AB through B to the lower edge of the rafter to locate point C. A line connecting points A and C will then show the miter cut.

D. The Snub Hip Rafter

The run of the snub hip rafter will be one-half the span of the gable plate (see Fig. 19-1). The length of this rafter in steps of the square may be found by subtracting the number of steps on the shortened common rafter (9) from the number of steps on the full common rafter (12). This will leave 3 steps of 8 on 17 for the snub common rafter. The ridge cut is a single cheek cut. This rafter should be shortened as described in Unit 16.

E. Jack Rafters

The length of the jack rafters for the snub gable is found by using the main common rafter as a pattern. In the plan (Fig. 19-1) the main common rafter has 12 steps of 8 on 12. The shortened common rafter has 9 steps. Therefore, if a common rafter were used from the snub gable plate to the ridge, the length would be the difference between 9 and 12, or 3 steps.

In this case, the jacks of the snub gable are spaced one foot each side of the ridge. This would be one step down from the ridge. Therefore, the length of these jacks would be 3 steps minus 1 step, or 2 steps of 8 on 12.

The length of the opposite jacks that run from the main plate to the snub hip would be one step less than the common rafter, or 11 steps.

How to Erect a Snub Gable

1. Lay out and cut the rafters as described.

2. Space the ridgeboard as usual, marking the location of the snub hip rafters the same distance from the end of the building as there are feet or inches of run in the snub hip rafter. See plan.

3. Lay out and cut the snub gable plate and shortened main common rafters.

4. Nail the snub gable plate and the shortened common rafters together at the mitered joints.

    NOTE: This process is best accomplished by nailing the plate and both rafters together on the subfloor and raising the plate and rafters as a unit.

5. Nail the common rafters to the plates at the seat cuts. Plumb and brace the rafters to the floor as in Unit 15.

6. Erect the snub gable hip rafters and jacks as described in Unit 16.

7. Fill in the main roof jacks which run from the snub hip rafters down to the main roof plate.

DORMERS BUILT ON THE ROOF

Small dormers without windows used as decorative features are often built on the surface of the roof after it has been sheathed. Larger dormers with windows may also be built on the roof, using a header as for the shed roof. However, the cuts on the rafters that fit on the main roof surface differ from the seat cuts of other rafters.

A. Rafter Cuts When Dormer and Main Roof Are of Equal Pitch

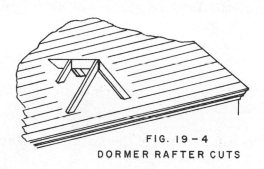

FIG. 19 – 4
DORMER RAFTER CUTS

The bottom cut of a dormer rafter which rests against the main roof sheathing may be described as a level cheek cut.

This cut is obtained by laying out the level or seat cut of the main common rafter on a piece of dormer rafter stock, Fig. 19-5. A distance equal to the thickness of the rafter stock (1 1/2") is measured from the top edge of the rafter along this level line. A line is then squared from the top edge of the rafter down to this point (A). From the top of this line, a line is then squared across the top edge of the rafter to point B. The cheek cut line is then drawn from point B to point C.

The lengths of the jack rafters are found from the common rafter in the same way as the jacks of the dormer, described in Unit 16.

B. Rafter Cuts When Dormer and Main Roof are of Unequal Pitch

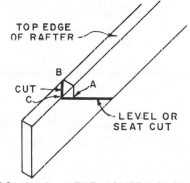

FIG. 19 – 5   LEVEL CHEEK CUT

The general shape of the cheek cut for this type of dormer rafter is similar to that shown in Fig. 19-5, but the layout is slightly different. To lay out this cheek cut, it is necessary to determine how much the main roof rises in a level distance equal to the thickness of the dormer rafter. Assume that the rise of the main roof is 9" and the run is 12". On a horizontal distance of 2" (the approximate thickness of a rafter), the rise would be 1/6 of 9", or 1 1/2". This distance is referred to as D, Fig. 19-6.

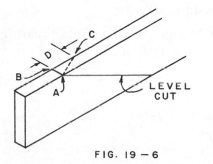

FIG. 19 – 6

The level-cut line of the dormer rafter is laid out using the rise and run of the dormer rafter. From the point where the level line meets the top edge of the rafter (A), a line is squared across the top of the rafter to point B. The distance D, which was found to be 1 1/2" in this case, is then measured from point B to obtain point C. From point C a line is drawn to point A to give the cheek cut for the bottom end of the rafter. The lengths and ridge cuts of the dormer rafters are found as described in Units 16 and 17.

How to Build a Small Dormer on the Main Roof

NOTE: In erecting small dormers on the sheathing of the main roof, it may be more convenient to place the longest dormer rafters on the main roof first by tacking them temporarily to the roof and to the dormer ridgeboard.

1. Cut the ridgeboard so that it fits against the main roof in a level position. This cut is the same as the bottom cheek cut of the dormer rafters. Make the ridgeboard long enough so that it will extend beyond the face of the longest rafters of the dormer.

2. Lay out and cut the jack rafters.

3. Tack the long jacks to both sides of the ridgeboard, keeping the tops of the rafters even with the top of the ridgeboard.

4. Level the ridgeboard from the top of the rafters back to the main roof. Tack it to the roof at this point.

5. Plumb the long jacks from the main roof up to the ridgeboard.

6. Recheck the ridgeboard and rafters for plumbness and nail the rafters.

7. Fill in the remaining dormer jacks that extend from the ridge to the main roof.

   NOTE: Be sure the ridgeboard is not forced out of line when nailing the jacks to the ridge and main roof.

REVIEW PROBLEMS   Unit 19

1. Describe an odd-pitch roof.

2. Which scale on the framing square would be used when laying out an odd-pitch rafter?

3. Describe the method used in laying out an odd-pitch rafter.

4. How would you find the true length per foot of run for an odd-pitch rafter when the total rise and run are greater than the length of the square?

5. Is the spacing and placement of the rafters for an odd-pitch roof different from an even-pitch roof?

6. Describe a snub gable.

7. Where is the plate for a snub gable placed?

8. What determines the size of the snub gable?

9. How long is the run of the snub hip rafter?

10. How is the length of the jack rafters for a snub gable found?

11. Describe the process of erecting the snub gable plate and shortened common rafters.

12. List the tools necessary to frame and erect odd-pitch rafters, snub gables, and small dormers.

Unit 20   BAYS AND OVERHANGS

The procedure used in framing overhangs for floors or flat roofs is very much like that used in the construction of bay windows and will be explained in this unit.

A bay window is a projection built on the outside wall of a building. It is designed to add more floor space to a room, to admit more light, or to improve the appearance of the building. The many styles and shapes of bay windows may be classified as square, three-sided, or circular. A description of one type of straight-line bay window will illustrate the basic principles of this type. The framing of the projection into the main floor joists of the building is more or less common to all types. The side-wall and roof construction will be described in terms of the more commonly used types.

FLAT-ROOF OVERHANG

In flat-roof construction, when an overhang is desired on all sides of the house, lookout rafters are installed. A double header is used, and the lookout rafters are fastened to the header and toenailed to the wall plate. The double header is usually placed twice the width of the overhang from the wall plate. The outside ends of the rafters are usually finished with some type of header as a nailing surface for whatever trim is to be applied. In plank-and-beam construction the beams are framed in much the same manner, and the planking and beams are left exposed on the underside.

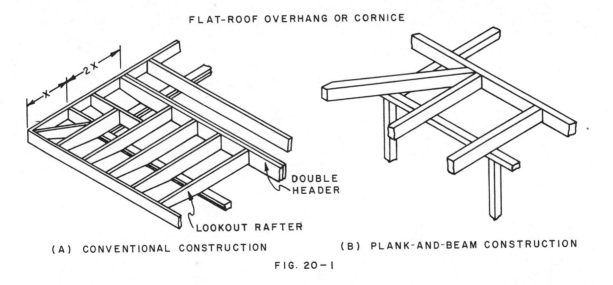

FLAT-ROOF OVERHANG OR CORNICE

(A) CONVENTIONAL CONSTRUCTION   (B) PLANK-AND-BEAM CONSTRUCTION

FIG. 20-1

FLOOR OVERHANG

It is sometimes desirable to have a projection or overhang of the second floor in a two-story house. The purpose of this projection may be to add more floor space to the second floor, to enhance the architectural appearance, or to cap off a brick veneer facing of the first floor. The overhang usually would appear only on the front of the house, and the joists would project to support the wall above. If it is required to parallel the joists, the same type of framing as shown for the flat-roof overhang could be installed (Fig. A-20-1).

BAY WINDOW FRAMING DETAILS

FLOOR FRAMING

When the floor joists of the bay window are to run parallel to the joists of the main building, the framing is rather simple. The joists are allowed to project beyond the outer face of the foundation wall the same distance the end wall of the bay extends beyond the main wall of the building.

The floor plan of the bay is laid out on the tops of the floor joists. The joists are cut along these marks, and a header is fastened to them to form a support for the subfloor, upon which is nailed the sole plate for the wall studs of the bay.

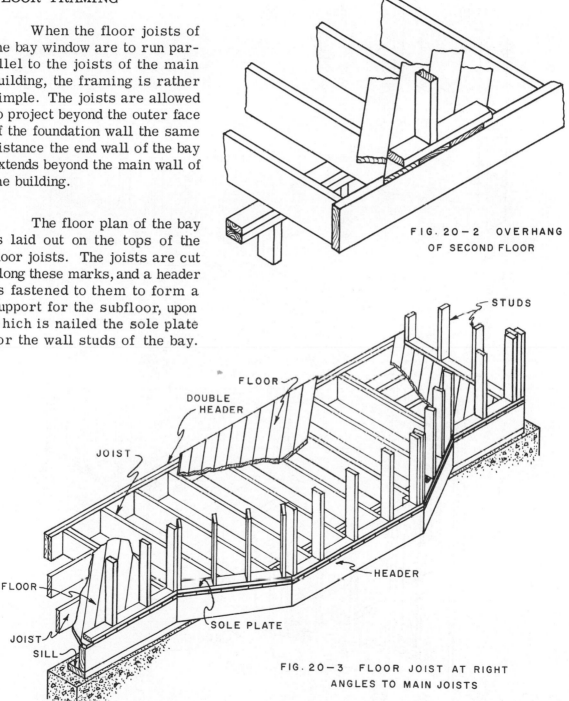

FIG. 20-2 OVERHANG
OF SECOND FLOOR

STUDS

FLOOR

DOUBLE
HEADER

JOIST

FLOOR

HEADER

JOIST

SILL

SOLE PLATE

FIG. 20-3 FLOOR JOIST AT RIGHT
ANGLES TO MAIN JOISTS

When the joists of the bay run at right angles to the main joists, it is necessary to use headers and trimmers in the main joists. This type of floor construction is necessary for the proper support of the outside walls of the bay.

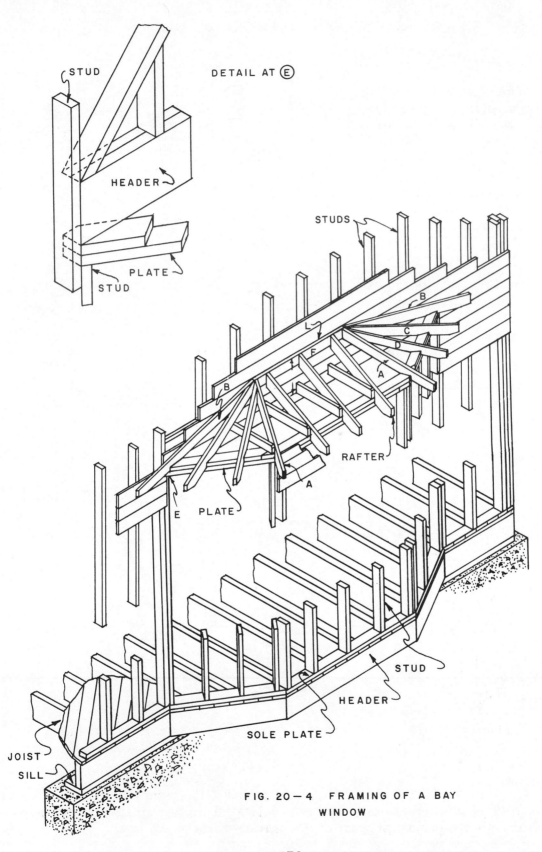

DETAIL AT Ⓔ

STUD

HEADER

PLATE

STUD

STUDS

B

C

D

L

F

A

B

RAFTER

E

PLATE

A

STUD

HEADER

SOLE PLATE

JOIST

SILL

FIG. 20—4   FRAMING OF A BAY
WINDOW

WALL FRAMING

The sole plate and the double plate at the top are laid out and spaced for the studs in the same way as the main walls of the building, Fig. 20-4. The double plate is lapped at the angles of the bay and is attached to the main wall as shown at E. The same length of stud may be used for the bay window as for the main walls. The studs at the angles of the bay are turned so that the edges are parallel to the outside edges of the upper and lower plates of the bay. The plates and studs are nailed together and raised in the same manner as other bearing partitions of the building. Refer to Unit 10. Provision for window openings is the same as described in Unit 11.

ROOF FRAMING

The roof shown in Fig. 20-5 is typical of roofs on bay windows. Each of the pairs of rafters at the ends must be laid out as individual hip rafters because they have different runs.

The distance from the face of the 2×4 (F) on the main building to the sheathing line at the outside of the bay window is shown at E. This distance is the run of the rafter. It may be assumed to be 20″ in this case. The vertical distance from the top of the double plate to the top of the 2×4 (F) is the rise of the rafter. This is 11″. In small roofs it is often more convenient to change the rise and run to inches.

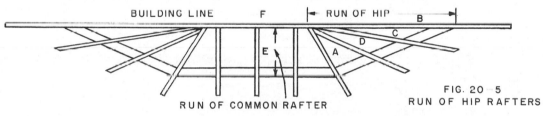

FIG. 20-5
RUN OF HIP RAFTERS

The common rafter is laid out by locating a measuring line on the rafter as described in Unit 15. A 2×4 is sufficiently strong for these rafters, and the measuring line should be at least 2 1/2″ below the top edge. This rafter is laid out with one step of the square. The figures 20 on the body and 11 on the tongue should be placed on the measuring line. A mark along the outside of the tongue will give the plumb cut, and one along the body will give the seat cut.

For the hip rafter at A, the rise will also be 11″. To find the run of this rafter, a full-size plan of the roof should be made on the subfloor. The run can be measured on this plan. The run of this rafter most likely will be greater than 24″ and, therefore, greater than the length of the square. Assuming that the run of rafter A is 26″ and that the rise is 11″, it will be necessary to divide these figures by 2. Two steps of 13″ and 5 1/2″ should be taken to find the length and cuts of the hip rafter. The cheek cuts are found from the full-size roof plan. The bevels are transferred from the drawing to the rafter by using a sliding T bevel.

The lengths and cuts of the other hip rafters (B, C, and D) are found in a similar manner, the run being measured from the spaced position of the rafter on the plate. The backing of the hip rafters, in this case, is negligible and may be omitted.

The rafters for a square or rectangular bay are laid out in about the same manner as those of a main roof.

TOOLS AND EQUIPMENT FOR FRAMING BAY WINDOWS

Hammer	Steel square
Crosscut saw	Rule
Spirit level	Sliding T bevel
Straightedge	

How to Frame Bay Window Floor Joists

1. Determine the location and size of the bay from the blueprint.

2. Place the floor joists in the usual manner, but allow them to project beyond the foundation line as described.

3. Mark the shape and angle of the bay on the tops of the joists.

4. Square the lines down along the sides of the joists and cut them on these lines. NOTE: When a box-type sill is used on the main building, the vertical member of the sill should be run around the bay.

How to Frame Bay Window Studding, Plates, and Openings

1. Cut pieces of 2×4 for the sole plate to follow the outline of the ends of the joists.

2. Cut the studs of the bay the required length.

3. Lay out and cut the top plates so that they will lap over each other at the corners.

4. Frame in the window openings in the same manner as for the other window openings of the main building. Refer to Unit 11. Apply the sheathing to the side walls of the bay.

How to Frame a Bay Window Roof

1. Lay out and cut the required number of rafters as described. The hip rafters should be cut in pairs.

2. Mark a level line along the face of the main building showing the height of the common rafters, see L, Fig. 20-4 on Page 176.

3. Place the hip rafters B against the building so that the tops of the rafters are even with the level line and so the seat cuts fit over the plate. Tack these rafters to the sheathing of the main building.

4. Place the 2 × 4 (F) against the sheathing of the building with the top edge in line with the level line. Its length should be the distance between the plumb cuts of the two hip rafters (B). Spike the 2 × 4 to the studs of the building.

5. Place the hip rafters (A) even with the top edge of the 2 × 4 (F). Place the seat cuts at the intersection of the side and front plates of the bay. Tack these rafters in position temporarily.

6. Place rafters C and D in a like manner.

7. Place the common rafters from the front plate to the 2 × 4 (F) on the building. These rafters should be spaced about 16″ o.c.

8. Check the tops of the hip and common rafters with a straightedge to see if they are even. If so, nail the rafters solidly.

REVIEW PROBLEMS   Unit 20

1. Explain the three functions of a bay window.

2. List three shapes of bay windows.

3. How far in from the edge of the wall plate is the double header placed for lookout rafters?

4. Why is some type of header placed on the outside ends of the lookout rafters?

5. Explain three functions of a floor overhang.

6. Where are floor overhangs usually constructed on a house?

7. Describe the joist framing for a bay or floor overhang when the joists are to run parallel to the main floor joists.

8. When is it necessary to use headers and trimmers in main floor joists when framing a bay or floor overhang?

9. Are studs of the same length used for the bay window as for the walls of the main building?

10. Why are the rafters at the ends of a bay window laid out as individual hip rafters?

11. Describe the run of a common rafter for a bay window.

12. Where should the measuring line be placed on the common rafter of a bay window?

13. How can you find the run of hip rafters of a bay window?

14. List the tools necessary to frame floor overhangs and bay windows.

Unit 21   PARTITION FRAMING

Due to the growth in popularity of basementless houses and low-pitched roofs, there is a greater demand for storage space on the main floors of houses. This demand is being satisfied by various types of studless thin wall panels which are being constructed so as to require less floor space and can be arranged as storage closets or room separators. The use of roof trusses and plank-and-beam framing overcomes the need for the partitions to carry any of the roof load. This makes it possible to use lightweight box-type framework which gives added strength to thin wall panels when used as partitions. The partitions may be built on the job from stock materials, or they may be prefabricated cabinet-type units.

The conventional type of partitions are still being installed to a great extent and can be used as a basis for the construction of job-built thin wall partitions. The plate layout and stud spacing for conventional partitions is about the same as the exterior walls of the house. This is especially true of the western platform type of construction, where the outside walls and the interior walls for bearing partitions are laid out exactly alike.

It is important that measurements be taken carefully from blueprints of floor plans. These plans should show the dimensions of rooms from the center of one wall partition to the center of the opposite partition. They should also show the thickness of the rough or finished partition.

The location and size of door openings and of all other openings should be carefully laid out on the plates of the partition when the stud spacing is marked. This is particularly true in the layout of bearing partitions.

BEARING PARTITIONS

Interior load-bearing partitions support the floor joists above them. They help to form a continuous bearing from the girder up through the building and also make the frame of the building more rigid. These partitions should be built in the same manner as the side walls and partitions in the platform frame (Unit 10).

Openings in these partitions for arches, doors, or supply ducts often interfere with the regular spacing of the studs. Such openings should be properly trussed and supported so the partitions will not be weakened. See Unit 11.

Studs of bearing partitions should be spaced the same distance apart as the joists above. If the studs are to be notched for horizontal pipes, the notches should be no deeper than one-third the depth of the stud. If they must be made deeper, a header should be installed.

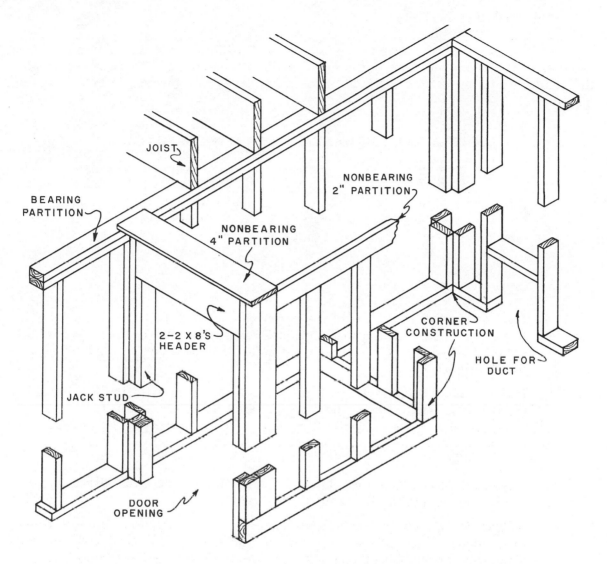

FIG. 2I—I INTERIOR PARTITIONS

NONBEARING PARTITIONS

Nonbearing interior partitions carry no load from the upper parts of the building. They are used to divide the floor space into rooms and to provide a surface on which trim and wall coverings can be fastened. They also supply support and enclosure for heating and plumbing lines and ducts. This type of partition may run parallel to or at an angle to the floor joists.

The studs of partitions around clothes closets are turned so that the 3 1/2″ side of the stud is parallel to the length of the partition. This makes the framing of the partition only 1 1/2″ thick. This construction is not recommended if the studs are over 81 inches long or if there is a door opening wider than 2′-6″ in the partition.

OPENINGS IN PARTITIONS

Openings for doors in partitions are framed with trimmer and jack studs. The jack studs run from the shoe to the under side of the header. These studs support the header and also stiffen the stud at the side. They also provide a better nailing surface for the interior trim.

TOOLS AND EQUIPMENT FOR PARTITION FRAMING

Steel square	Steel tape
Hammer	Spirit level
Saw	Chalk line
Straightedge	Compass saw

How to Frame Bearing Partitions

1. Lay out the exact location of the partitions on the subfloor. If the partition is long, it is advisable to snap a chalked line showing one edge of the partition location. If the partition is short, the straightedge may be used as a guide for the line.

   NOTE: The cross partition lines should be square with the side walls of the building. Both ends of the parallel partition should be the same distance from the side walls. Check all corners of the room layout to see that they are square.

2. Lay out the top and bottom plates of the partition. Mark the rough openings for the doors and the locations of the studs.

   NOTE: Make the rough door openings 2 1/2″ wider than the finish mill size of the door. This allowance is for the thickness of the two side jambs and allowance for plumbing the jambs. The allowance for the head jamb is 1 1/2″ and for the finish floor, 1″.

   The window rod used to mark the height of window headers in laying out window openings in side walls may be used to mark the door headers in partitions. Refer to Unit 11.

3. Cut the top plate to the proper length.

4. Cut the bottom plate to the correct length. It may be cut in two pieces, one for each side of the door, Fig. 21-2.

   NOTE: Some carpenters prefer to leave the top and bottom plates the full length of the partition. The plates are then spiked to the tops and bottoms of the studs, the partition is raised, and the bottom plate at the door opening is later cut out. Refer to Unit 10.

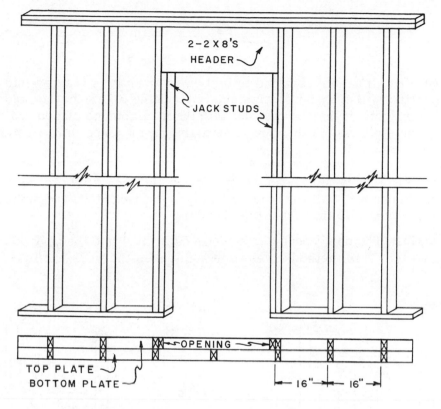

FIG. 21-2 PARTITION LAYOUT

5. Nail the bottom plates to the subfloor, keeping them in line with the chalk line snapped on the floor and spacing them correctly at the door-opening marks. Spike the plates to the joists with 16d spikes.

6. Lay the top plate on the bottom plate and with an extension measuring stick measure the distance from the top of the two plates to the bottom of the joists above. This distance is the length of the studs.

   NOTE: If a double plate is to be used at the top of the partition, lay an extra 2 × 4 block on the plates when taking the measurement for the studs.

7. Cut the required number of studs and spike them to the upper plate at the stud marks.

8. Raise the partition and toenail each stud with four 8d nails to the bottom plate at the stud marks.

9. Plumb the partition both ways at each end and in the middle, using a straight-edge and a spirit level.

10. When the partition is straight, spike the top plate to the undersides of the joists above.

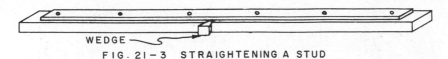

FIG. 21-3 STRAIGHTENING A STUD

NOTE: If a stud is found to be bowed after it is framed into the erected partition, it may be straightened by cutting a kerf in the stud and forcing it straight. It may be checked with the straightedge placed against the edge of the stud. When the stud is straight, nail a piece of 3/4" stock along the side to hold it in place.

11. Frame in the door headers by nailing the jack studs to the studs at each door opening. Nail the headers on top of the jack studs and spike through the side studs into the ends of the header.

NOTE: The door opening is sometimes framed into the partition before it is raised. This method saves time and makes the work easier.

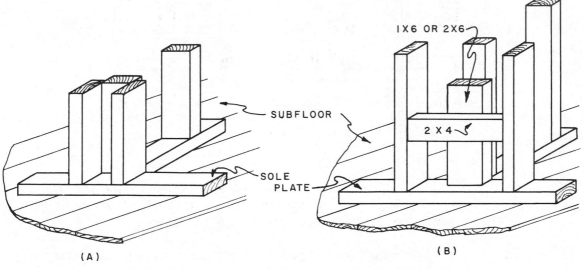

FIG. 21-4 INTERIOR WALL INTERSECTION

How to Frame Nonbearing Partitions

NOTE: The framing and erection of this type of partition is similar to that of the bearing partition. Since the non-bearing partition generally runs parallel to the joists, provision must be made for nailing the top of the partition between the joists.

1. Cut pieces of 2 × 4 to fit between the ceiling joists. Space the 2 × 4 blocks about 4' apart. Keep the blocks up the thickness of the lath nailer strip above the bottom edge of the joists.

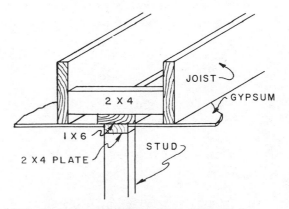

FIG. 21-5 SECURING THE TOP OF A PARTITION PARALLEL TO JOISTS

2. Cut a piece of 1 × 6 as long as the partition for a lath nailing strip.

3. Gage a line along the face of the lath nailer about 3/4″ from the edge. This will allow the nailing strip to project beyond each side of the top plate and will provide backing for the ceiling lath.

NOTE: Lay out the partition location and cut the studs in the same manner as for a bearing partition except that a single top plate may be used with the lath nailing strip.

4. Assemble, raise, plumb, and fasten the partition in the same manner as explained for the bearing partition.

How to Make Openings for Heating Pipes

1. Mark the size of the heating duct on the plate of the partition where the duct is to be placed. Make the duct opening large enough to allow at least 1/2″ on all four sides so that the duct does not come in contact with the wood framework.

2. Cut the plate out between the studs. If possible, make the cuts 1/2″ away from the studs to avoid hitting the nails with the saw.

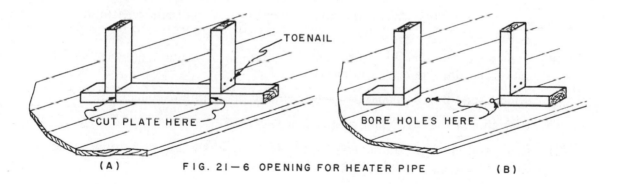

(A)   FIG. 21—6 OPENING FOR HEATER PIPE   (B)

3. Bore two holes through the subfloor diagonally opposite each other.

4. Insert a compass saw in the holes and cut out the subfloor to the proper size to admit the heating duct.

REVIEW PROBLEMS   Unit 21

1. Why is there a growing demand for storage space on the main floors of houses?

2. Studless thin wall panels are often used to form storage closets or room separators. Why?

3. State two factors which overcome the need for the partitions to carry any of the roof load.

4. Does lightweight box-type framework give added strength to thin wall panels?

5. Can thin wall partitions be built from stock materials, or must they be prefabricated cabinet-type units?

6. Are conventional partitions still being installed to any extent?

7. Is the stud spacing and plate layout of conventional partitions the same as for the exterior walls of the house?

8. What is the purpose of a load-bearing partition?

9. How can openings in partitions such as arches or doors be installed?

10. How far apart should studs for bearing partitions be placed?

11. How deep may notches be cut in a stud if no header is installed?

12. What is the purpose of nonbearing partitions?

13. May nonbearing partitions run parallel to the floor joists?

14. How are the studs placed in a thin wall nonbearing conventional partition?

15. What is a jack stud and where is it used?

16. How much wider should the rough door opening be than the finish mill size of the door?

17. How much higher should the rough door opening be than the finish mill size of the door?

18. Can the window rod used to mark the height of window headers be used to mark the door headers in partitions?

19. Which size of nail should be used to fasten the plates to the joists?

20. How should you straighten a stud that is bowed?

21. Draw and label a sketch showing the framing at the top of a nonbearing partition running parallel to the joists.

22. How much larger than a heating duct should the opening for it be?

23. List the tools necessary for framing partitions.

Unit 22   FURRING, GROUNDS, AND BACKING

The carpenter should have a general understanding of the methods used by the other building tradesmen to fasten their various fixtures to the frame of the building. He should know how to read from the building drawings the locations of plumbing, electrical, and heating fixtures which are to be installed. This information is necessary so that he may provide proper backing to which these fixtures may be attached after the lath and plaster have been applied to the walls.

Furring strips, grounds, and backing must be installed before the lath is applied. They should be of regular framing material and should be free of knots and cracks. They must be firmly secured so that they will not become loosened when the nails or screws holding the fixtures or trim enter them.

FURRING STRIPS

Furring strips at least 3/4″ by 1 1/2″ are required on masonry walls less than 12″ in thickness in habitable spaces. The spacing and thickness of the strips depends upon the type of insulation or the interior finish to be used. They are applied to a surface to support plaster, stucco, or other surfacing material. They may be used to form an even surface over a rough or irregular wall or to provide an air space between the inner surface of the outside wall and the finish surface of plaster or other finishing material. This space acts as an insulator and prevents condensation. If additional insulation is desired, it may be placed between the furring strips. Furring strips also help to prevent plaster cracks due to the difference in the rate of settling between the masonry and the framework of the building.

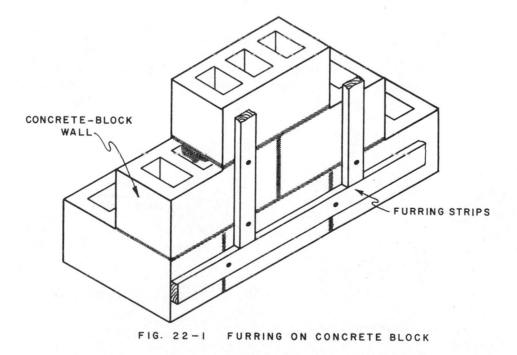

CONCRETE-BLOCK WALL

FURRING STRIPS

FIG. 22-1   FURRING ON CONCRETE BLOCK

PLASTER GROUNDS

Plaster grounds are strips of dimensioned stock, generally the thickness of the combined lath and plaster coats. Their function is to form a straight rigid surface to which the plasterer may bring the finish coat of plaster and to which the carpenter may secure the interior trim. Plaster grounds should be installed at openings where the frames do not provide a ground and at the intersection of walls with the floor.

Two types of plaster grounds may be used, those that remain in place and those that can be removed after the plastering has been completed. The removable type are sometimes used around interior door openings and consist of narrow strips that are nailed to the faces of the studs and header and spaced to the width of the door jamb to form a good casing fit when the trim is installed. The grounds are removed after the plaster is dry.

Another method is to install the finish jamb before plastering and leave it in place. This can be done only when the trim of the building is to be painted as the jamb must be painted before it is installed to protect it against moisture.

BACKING

Backing may be of any dimensioned stock, but 1×6 or 1×8 is generally used. The function of backing is to provide a solid surface to which the lath, plasterboard, or fixtures may be attached. Backing is necessary at the intersections of partition walls and ceilings. It is also required where a solid base is needed to hang such fixtures as sinks, bathroom and lighting fixtures, heating fixtures, and special built-in sections.

CORNER BEAD

Corner bead and metal arches are metal ribs which are nailed to the studs at exposed corners which are to be plastered. Their function is to protect the plastered corner from becoming chipped and to form an even contour for plastered openings.

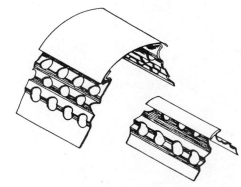

FIG. 22—2 CORNER BEADS

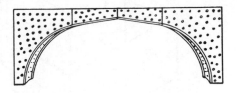

FIG. 22—3 METAL ARCH

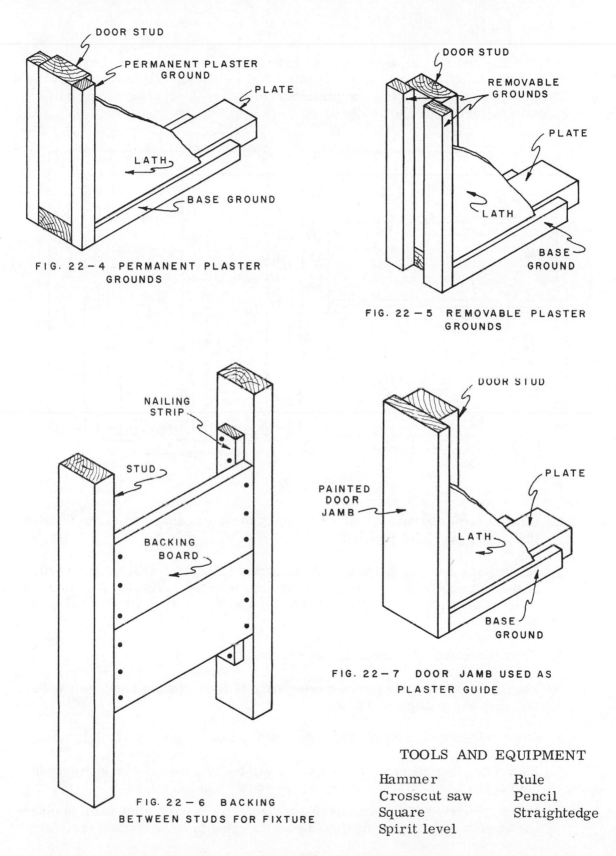

DOOR STUD

PERMANENT PLASTER
GROUND

PLATE

LATH

BASE GROUND

FIG. 22−4   PERMANENT PLASTER
GROUNDS

DOOR STUD

REMOVABLE
GROUNDS

PLATE

LATH

BASE
GROUND

FIG. 22−5   REMOVABLE PLASTER
GROUNDS

NAILING
STRIP

STUD

BACKING
BOARD

FIG. 22−6   BACKING
BETWEEN STUDS FOR FIXTURE

DOOR STUD

PLATE

PAINTED
DOOR
JAMB

LATH

BASE
GROUND

FIG. 22−7   DOOR JAMB USED AS
PLASTER GUIDE

TOOLS AND EQUIPMENT

Hammer	Rule
Crosscut saw	Pencil
Square	Straightedge
Spirit level	

How to Apply Furring Strips

NOTE: Furring strips may be applied to masonry walls either horizontally or vertically, whichever way is best suited for the finish wall covering. If the finish wall material is to be applied horizontally, the furring strips will be placed vertically, 16″ on centers.

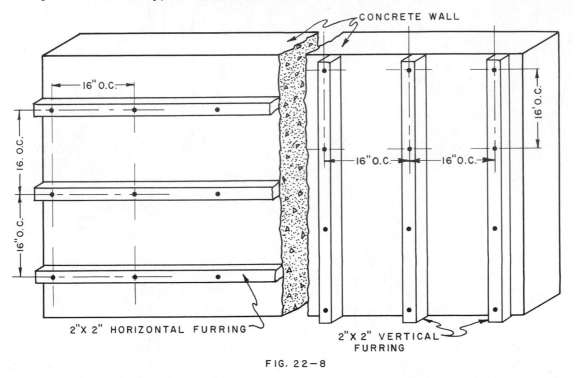

FIG. 22—8

1. Cut the required number of strips long enough to reach from the floor to the ceiling (vertical furring).

2. Starting at a corner, fasten the furring strips to the wall with hardened nails; or the wall may be drilled and soft wood plugs inserted to receive the common nails. There are many patented fasteners on the market which also might be used.

3. The nails should be placed 16″ on centers.

4. Usually every third or fourth furring strip is first installed and fastened so that they are straight and plumb.

5. Wood shingles may be used to shim the furring strips to a straight line.

6. Fasten the other strips in between and straighten with the aid of a straightedge held against the outer faces of the previously applied strips.

NOTE: Furring strips are applied horizontally to wood stud walls in much the same manner, when the finish wall covering is to be applied vertically.

How to Install Plaster Grounds

NOTE: Base grounds may be placed at the floor line only if gypsum lath is to be applied. If expanded metal lath is to be applied, a double ground is used.

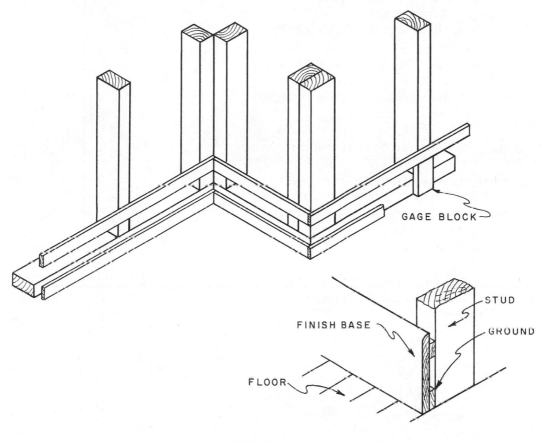

FIG. 22-9

1. Nail the top ground strip to each stud 1″ below the point where the top of the finish baseboard will be. Use 6d nails.

    NOTE: Use a gage block placed between the bottom of the ground strip and the floor. This will avoid the necessity of repeated measurements to get the ground straight.

2. Check the face of the ground strip with a straightedge to see that it is straight throughout its entire length. If there are any irregularities, use a piece of wood shingle to wedge between the stud and the back of the ground. Drive the wedge down until the face of the strip is straight.

3. Nail the bottom ground strip to the studs, keeping the strip tight against the top of the subfloor.

How to Prepare Grounds around Door Openings

1. Nail the ground strips around the edges of the door to the vertical studs and the header.

    NOTE: Use a straightedge if necessary to straighten the faces of the strips and to maintain a uniform width between these faces on each side of the opening. A piece of uniform width similar to that shown may be used in plumbing and keeping the strip straight and uniform.

How to Install Backing for Partition Intersections

1. When laying out the partition location on the subfloor, mark the positions of the studs A-A which are to be used as backing. These studs should be about 2″ apart.

2. Erect and nail these studs in position when the other partition studs are nailed.

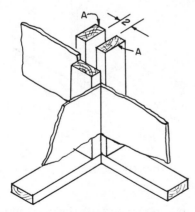

FIG. 22—10 BACKING FOR LATH AT PARTITION INTERSECTION

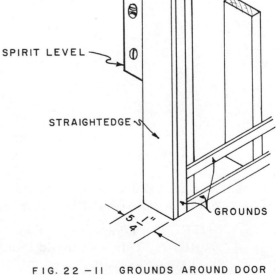

FIG. 22—11 GROUNDS AROUND DOOR OPENING

How to Fur Grounds and Backing

NOTE: Another method may be used to back the lath at wall intersections.

1. Nail pieces of 2 × 4 between the studs B-B of the partition which has already been erected. These pieces should be inserted between the studs every 4′ in the position shown. The face of the 2 × 4 should be 1″ from the faces of the studs B-B.

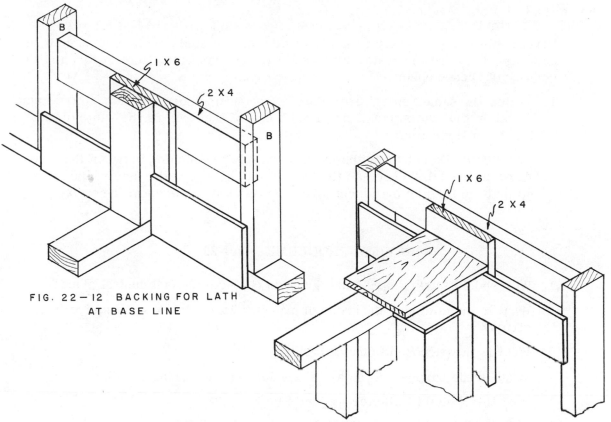

FIG. 22—12 BACKING FOR LATH
AT BASE LINE

FIG. 22—13 BACKING AT CEILING LINE

2. Nail a piece of 1 × 6 on the back of the corner stud of the partition to be erected so that it projects about 1″ beyond each side of the stud. The 1 × 6 should be nailed to the cross 2 × 4 pieces after the partition has been raised and plumbed.

NOTE: The same method used at the intersection of walls may be used at the intersection of the ceiling and a side wall. The backing board need be only 1″ thick. The process of installing this backing is explained in Unit 21.

How to Install Corner Bead

1. Nail the corner bead temporarily to the stud corners at the top of the opening.

2. Adjust the corner bead so that the finished face projects beyond the corner of the stud the thickness of the plaster coats.

3. Use a straightedge and spirit level to plumb and straighten the metal bead.

4. Nail the bead to the stud with shingle nails every 6″ of its entire length.

How to Install Metal Arches

NOTE: Metal arches may be made to fit over the studs of the partition at the opening. They are in several styles and save the carpenter time in forming the contour of arched openings.

1. Follow the same general procedure as for setting corner beads. Care must be taken that the spring line at each end of the arch is the same distance from the floor line.

    CAUTION: Do not force the metal arch into the opening between the studs. Allow at least 1/2" between the back of the metal and the face of the header so that the plaster may form a key through the openings in the metal arch.

REVIEW PROBLEMS   Unit 22

1. Why should furring strips and grounds be securely fastened in place?

2. What is the minimum thickness of masonry walls that do not require furring strips?

3. What is the purpose of furring strips?

4. What is the purpose of the air space formed by furring strips?

5. How thick should plaster grounds be?

6. State two functions of plaster grounds.

7. Where should plaster grounds be placed?

8. Describe two common types of plaster grounds.

9. When can the finished jamb be installed as a plaster guide, and what precaution should be taken?

10. What is the function of backing?

11. List four fixtures which usually require backing.

12. What is the function of corner beads and metal arches?

13. When is it necessary to apply furring strips horizontally?

14. How far apart should furring strips be placed?

15. State two ways in which furring strips may be fastened to a concrete wall.

16. How far apart should the nails be placed in furring strips?

17. When is it necessary to apply a double base ground?

18. How far below the top of the finished base should the top of the ground be placed?

19. State the advantage of a gage block when applying base grounds.

20. How can grounds be straightened throughout their length?

21. How can a uniform width be maintained when applying permanent grounds around a doorway?

22. How far apart should the extra studs be placed when installing backing for partition intersections?

23. Draw a sketch of a partition intersection using a 1 × 6 as a nailing strip for lath.

24. How far should the corner bead project beyond the corner of a stud?

25. What type of nail should be used to fasten corner beads to studs, and how far apart should they be placed?

26. What is the spring line of an arch?

27. List the tools required to apply grounds, backing, and furring strips.

Unit 23   STAIR FRAMING

More care and knowledge is required in planning stairs, more ingenuity is necessary in laying them out, and more skillful workmanship is called for in their construction, than in any other work about the building. The most important consideration in designing stairs is to arrange them so as to afford the greatest ease of communication between the stories that they connect. Proper headroom should be provided so that any person walking up or down the stairs will have several inches of clear space above his head at all points. When possible, the stairs should be divided into two flights, as a long flight suggests considerable effort to walk up and down it. There is also a sense of danger produced when using such a flight, especially in the minds of infirm or very young people.

There are two types of stairs in a house, main stairs and service stairs. The main stairs are usually a special feature in the design of the house and afford ease and comfort in their use. The service stairs, leading to the attic or basement, are usually constructed of less expensive material and are somewhat steeper.

According to the Federal Housing Administration, the headroom must be continuous and measured vertically from the front edge of the tread to a line parallel with the stair run. This distance for main stairs is a minimum of 6'-8" and for basement and service stairs 6'-4". The width of the main stairs should be a minimum of 2'-8" clear of the handrail, and for service stairs 2'-6" clear. The minimum width of treads is 9" plus 1 1/8" nosing. The risers may have a maximum height of 8 1/4". A handrail must be installed on at least one side of each run on all stairs. The handrail should be 2'-8" above the tread at the riser line.

The framework of a staircase should be installed in accordance with the measurements of the finished stairs. Therefore, templets are generally laid out for the finished stairs at the time the framing of the rough stairs is installed. Measurements for both the rough and finish work are taken from the templet so that there will be no difficulty in fitting the finished stairs to the framing after it has been lathed and plastered.

FORMS OF STAIRS

There are many forms of stairways, but in this unit only the framing of the common types used in dwellings will be considered. The finish work of the stairs will be covered in a later book of this series on "Exterior and Interior Trim".

The method of framing the joists around a stair well is explained in Unit 8. More detailed information on the layout of templets for carriages will be given in this unit.

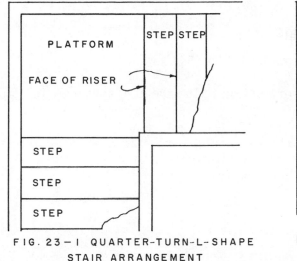

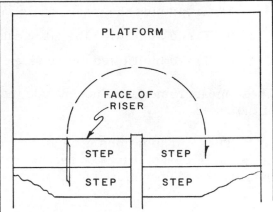

FIG. 23-1 QUARTER-TURN-L-SHAPE
STAIR ARRANGEMENT

FIG. 23-2 HALF-TURN OR U-SHAPE
STAIR ARRANGEMENT

The straight flight of stairs is often used when the stairs are located at the side of a room or between partitions that separate rooms. This type of stairway takes up considerable space but may be used to advantage where the space underneath is used for another straight flight or for closet space.

The quarter-turn or L flight, Fig. 23-1, is used where the stair space is somewhat restricted and where the platform may be located high enough from the floor to provide a doorway underneath. This space may then be used as a closet or passage into another room.

The half-turn or U flight, Fig. 23-2, is also used to conserve space. In both the L and the U types, a platform is necessary to afford passage at the point of change in the direction of travel on the stairs. This platform may be located midway between the two stories, thus making the rise and run of the two flights of stairs the same. It may also be placed above or below the midpoint, thus making the rise and run of the two flights different. The latter allows for a more flexible arrangement of the stairs in terms of the surrounding room arrangement.

PITCH OF STAIRS

One of the primary considerations in the layout of stairs is the safety of the people using them. There must also be ample space for the passage of furniture, both between the tops of the steps and the ceiling and also around the corners. The proper pitch of a stairway is one of the factors which helps to assure safety with adequate clearance.

The pitch of the stairs is dependent upon the proportion of the height of the riser to the width of the tread. There are several rules for the sizes and proportions of treads and risers. As the height of the riser is increased, the width of the tread must be decreased for comfortable results. A very satisfactory proportion will be found by any of the rules which follow.

1. Tread plus the riser equals 17″ to 18″.

2. Tread plus twice the riser equals 24″ to 25″.

3. Tread multiplied by the riser equals 70″ to 75″.

These measurements do not include the projection of the tread nosing over the face of the riser.

These proportions give an approximate pitch to the stairs of from 30 to 40 degrees. Although it may be necessary to change this angle, a steeper pitch becomes dangerous and uncomfortable, especially in a long flight.

DIMENSIONS OF STAIRS

Fig. 23-3 shows a cross section of a series of straight-run stairs such as might be used in a dwelling. These include the cellar stairs, the stairs from the first floor to the second floor, and the ones from the second floor to the attic.

The headroom is the vertical distance from the top of the tread at the riser line to the underside of the flight or ceiling above and is shown in the figure at H. Although it varies with the steepness of the stairs, it should be a minimum of 6′-4″ for service stairs and 6′-8″ for main stairs.

The clearance, which is necessary for the free passage of furniture, is shown at C and is the shortest distance between the top of the nosing and the underside of the flight above. This measurement is taken on a line perpendicular to the pitch of the stairs. The clearance at the approach and at the landings is shown at A. This should never be less than 3′-0″.

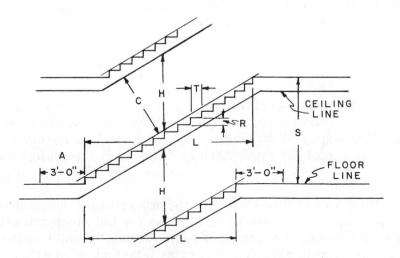

FIG. 23 - 3   DIMENSIONS FOR HEADROOM OF STAIRS

198

The total rise of the stairs is the distance from the top of the first floor to the top of the second floor and is shown at S. The run of the stairs is the distance from the face of the bottom riser of the stairs to the face of the top riser and is shown at L.

The number of treads and risers is found by dividing the total height of the story from the top of one floor to the top of the next floor by the height of the stair riser. The higher each riser is, the fewer risers and treads will be needed, and consequently the less space the stairs will take up. This is the reason why the width of the risers is often increased and the width of the treads decreased. This practice is permissible in some instances, but the rise should not exceed 8 1/4", and the tread should not be less than 9". It is very important that all risers be the same in any one flight of stairs.

Assume that the total height from the top of one floor to the top of the next floor is 8'-6", or 102", and that the riser is to be not over 8 1/4". 102 divided by 8 would give 12 3/4 risers. However, the number of risers must be a whole number. Since the nearest whole number is 13, it may be assumed that there are to be 13 risers.

102 divided by 13 equals 7.84, or 7 13/16", for the height of each riser.

After the height of the riser has been found, the width of each tread may be found from the rules of proportion previously stated. Rule one: Tread plus riser equals 17" to 18". Therefore, the tread equals 17 minus 7.84, or 9.16".

The total length of the stairs would be 9.16 (width of tread) multiplied by 12 (number of treads), which equals 109.92, or 9'-1 15/16".

If the total length of the stairs could be increased, the number of risers would increase, and the height of each riser would be less, thus making an easier stair to climb. In this case, if 14 risers were used, the height of each riser would be: 102 (total rise) divided by 14 equals 7.28, or 7 1/4". The tread would be 17 minus 7.28, which equals 9.72, or 9 3/4". The total length of the stairs would be 9.72 (width of tread) multiplied by 13 (number of treads) or , 10'-6 3/8".

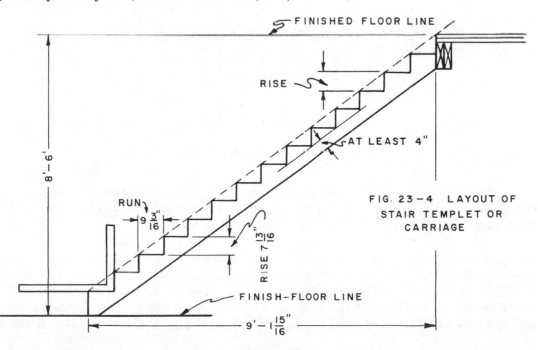

FIG. 23-4 LAYOUT OF STAIR TEMPLET OR CARRIAGE

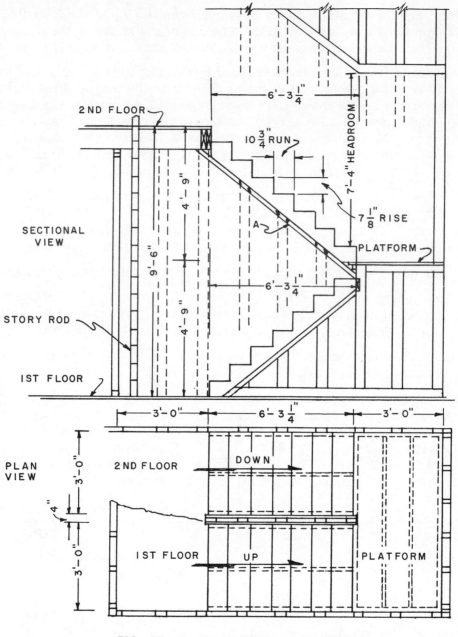

FIG. 23 — 5   A HALF—TURN STAIRWELL

Assume that a half-turn staircase is to be laid out. The total rise between the top of the first floor and the top of the second floor is 9'-6". The width of the platform is 3', and the run of each flight of stairs is 75 1/4".

A story rod is first laid out by marking the exact distance between the top of the first floor and the top of the second floor on a rod about 11' long. This distance is 114". A pair of dividers is set at a little over 7" and the rod is stepped off. If there is not an even number of steps, the setting of the dividers should be changed and the process repeated until the last division falls on the 114" mark. If the spacing is correctly performed, there should be 16 divisions of 7.13", or approximately 7 1/8". This distance is the height of each riser from top of tread to top of tread.

If the platform is to be mid-distant between the floors, the height from the top of the lower floor to the top of the platform is taken from the eighth division on the story rod. If the platform is to be placed above or below the mid-point, the top of the platform floor should line up with one of the divisions on the story rod. The flooring of the platform should be considered as a tread of the stairs.

The approaches at the first floor and the platform landing should each be at least 3' wide. The entire length of the stairway will be 75 1/4" plus 3' for each of the two landings making a total of 12'-3 1/4" between the two end partitions of the staircase.

LAYOUT AND PLACEMENT OF STRINGER

The stair templet or carriage of these stairs is laid out in about the same way as the templet of the straight-run stairs. Six steps are taken along a 2 × 10 plank, using 7 1/8" on the tongue of the square for the rise, and 10 3/4" on the body for run.

From the top tread mark on the stringer, a line is squared parallel to the riser marks. This line is shown as the top header cut at A and is the portion of the stringer which is fastened to the platform header.

The bottom stringer cut, which fits on top of the floor, is found by squaring from the last riser mark. This line is shown as the floor cut and is the dotted line at B. The solid line parallel to this dotted line shows the deduction to be made when the treads are to be nailed above the tread marks where the stringer is cut out. See the dotted lines showing the tread in position (Fig. 23-7).

When the finish stairs are to be built at a mill, temporary stairs are built of two or three cutout stringers which are nailed to the header of the platform. Temporary treads are nailed in place to provide a stairway for those working in the building. The stringers may also be used as a templet for the erection of the bridging of the stairs.

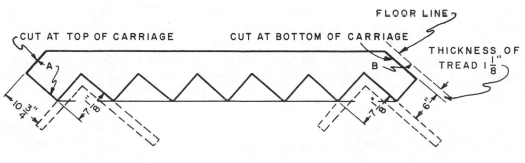

FIG. 23-6   LAYOUT OF STRINGER

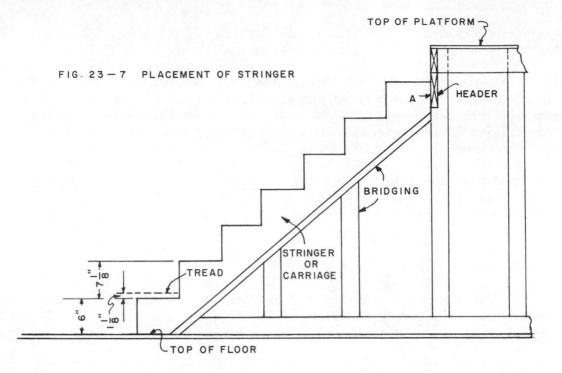

FIG. 23 — 7   PLACEMENT OF STRINGER

TOP OF PLATFORM

A →   HEADER

BRIDGING

STRINGER
OR
CARRIAGE

TREAD

$7\frac{1}{8}"$

$6"$ $5\frac{1}{8}"$

TOP OF FLOOR

DESCRIPTION OF STAIR BRIDGING

Stair bridging is the 2×4 framing that supports the staircase. The method of building the bridging depends upon the type of staircase. The 2×4 framing at each side of the staircase may completely enclose the stairs. The studs would extend from the floor to the ceiling.

The studs are shown extending from the floor to the bottom of the stairs only. The staircase is open from the top of the treads to the ceiling. This bridging is framed the same as a partition except that the top plate is run parallel to the slope of the stairs. The 2×4 furring at the underside of the stairs is shown in the sectional view (Fig. 23-5). This furring provides a base to which the lath and plaster may be secured, thus forming an enclosure for the underside of the stairs.

The partitions or bridging are often laid out, erected, lathed, and plastered before the stairs are constructed. The finish stairs are then built as a separate unit and fitted to the partitions or bridging when the rest of the interior finish is applied.

Sometimes the partitions and bridgework are erected and the stair carriages are installed so that the finish treads may be fitted to them after the walls have been plastered.

When the finish stairs are installed at the time the partitions are set, they should be thoroughly covered and protected until the interior trim is finished.

For any of these methods, stair templets should be laid out and the bridging constructed in accordance with them.

LAYOUT OF WINDERS

Generally the L and U types of stairs are built with a straight platform; but, where the stair space is further restricted, risers and treads are built upon the platform. These are called winders. (See Fig. 23-8.)

This practice is not recommended and should be avoided if possible. However, if winders are necessary, they can be built so they are reasonably safe.

The arrangement of the winders at the platform of the stairs is dependent upon the number of risers that are included around the turn. However, the fundamental principle is that the winders should be so placed that the width of the tread at the line of travel should be the same as the width of the treads throughout the entire staircase. The line of travel is taken to be 18″ from the inside edge of the internal stringer or handrail. Since each of the straight treads of these stairs is 9″ wide, each of the winders should be 9″ wide at this 18″ point.

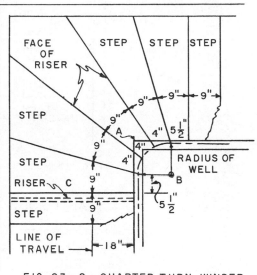

FIG. 23-8   QUARTER-TURN WINDER ARRANGEMENT

The lines of the risers should not converge to the point A, but rather to a point B, which lies beyond the intersection of the inside stringers of the stairs. The distance from B to the stringer is the radius of the stair cylinder. This does not necessarily mean that the intersection of the stair stringers at this point must be curved. It merely locates the center of the quarter circle and shows the quarter circumference which is divided by the number of risers on the platform. These divisions give the width of the treads at the inside stringers. See the 4″ spacings.

With a radius equal to the distance from point B to a point 18″ from the inside stringer, another arc is drawn using B as a center. The circumference of this arc is spaced in divisions equal to the width of the treads of the straight flight. See the 9″ spacings. The spacing is started from the face of the riser C of the straight flight. These points on the circumference are then connected by straight lines to the 4″ spacings on the small circumference. These straight lines give the locations and show the arrangement of the winders.

The arrangement of the winders in a U-shape stairway is shown in Fig. 23-9. The diameter of the U is given but the construction at this point could also be square, as shown by dotted lines. The circular lines merely illustrate the method of spacing the risers at the inside of the turn. In this type of stairs, only a certain number of risers may be included in the half turn. The horizontal distance between the two inside stringers would necessarily be 12″ in order to provide 10 1/2″ treads at the line of travel. The method of spacing the risers is similar to that of the L type of stairs.

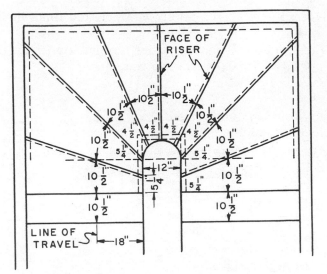

FIG. 23 — 9   HALF-TURN WINDER ARRANGEMENT

TOOLS AND EQUIPMENT FOR STAIR FRAMING

Hammer	Steel square
Crosscut saw	Sliding T bevel
Rule	Spirit level
Dividers	Story rod

How to Make and Install Carriages

1. Find the exact distance from the top of one floor level to the top of the floor above.

2. Lay off this distance on a story rod. This rod may be made of any smooth-faced board. A piece of finish flooring makes a good rod.

3. Space the required number of risers with dividers, equally, over the distance marked on the rod as explained.

    NOTE: If a single run is to be used with no platform between the floors, a templet may be laid out on a sheathing board. This is to be used as a guide for the erection of the bridgework.

    If stair carriages are to be used, they should be laid out on a 2×10 or wider plank so as to have at least 4″ of solid stock between the cutout section (for the risers and treads) and the edge of the plank.

4. Lay out and cut at least two planks for the carriages. The blocks cut out of one carriage may be nailed on the top of a 2×4 to make a third carriage. Be sure to deduct the thickness of the treads from the height of the bottom riser of the carriage as shown.

5. Space the three carriages as shown by dotted lines in Plan View (Fig. 23-5).

6. Install the three carriages against the header of the floor above as shown.

NOTE: Keep the side carriages about 1/2″ from the face of the partition studs if the stairs are to be enclosed. This is to allow a space for the lath to run between the outside face of the carriage and the inside face of the stair partition.

If the stairs are to be enclosed on both sides, it is best to erect the partitions before placing the stair carriages so that the carriages may be spiked to the studs for support.

7. Nail boards on the carriages at the tread cuts to provide temporary treads.

How to Build a Stair Platform

NOTE: In placing a platform in a stairway, the partitions that enclose the stairs at this point should be erected first. The location of the top of the platform may then be marked on the studs and the platform spiked in position when it is completed.

1. Determine the length and the width of the platform by taking the distance between the studs where the platform is to be located.

NOTE: If the platform is to be supported by the partition studs on all four sides, 2×4s may be used for building the platform. If the platform cannot be well supported by the studs, or if there is a span of over 6′ in the platform, 2×6s should be used in making the platform. The platform is framed as shown by the dotted lines in the Plan View (Fig. 23-5).

2. Build the framework of the platform to the proper size with 2×4s or 2×6s. Use 20d nails and space the middle members 16″ o.c.

3. Mark the height of the platform on the studs from the story rod.

4. Temporarily spike the platform in position at these marks.

CAUTION: Remember that the marks on the rod represent the tops of the treads and also represent the top of the finish floor of the platform.

5. Spike the platform to all the supporting partition studs when it is leveled and in its proper position.

6. Lay the flooring on the platform.

NOTE: The carriages for the platform type of stairs should be laid out, cut, and erected in the same manner as those of the straight stairs and as shown.

How to Erect the Bridging of Platform Stairs

NOTE: The bridging of platform-type stairs may be more difficult to install than that of the straight type because very often one of the enclosing partitions runs only from the first floor to the bottom of the stair stringer of the upper flight, as shown by dotted lines in the sectional view (Fig. 23-5). The bridging of the lower flight runs from the bottom of the lower stair carriage to the floor line. See the solid lines.

1. Lay out and erect the partition that runs from the bottom of the stair carriage of the lower flight to the floor.

    NOTE: If the rough stair carriage is to be permanent, the top plate of the partition may be built to support it. If the finish stairs are to be made in the mill and an open stringer is to be built at this time, the bridging should be 2″ below the bottom of the carriage or templet. This allowance is to provide clearance between the bottom back of the stair risers and the top of the stair bridging.

2. Set the stair carriages and partitions so that they are equally spaced at the platform level and at the floor line. This is important if an open stringer is to be used in the finish stairs. See the dotted lines showing carriage spacing in the Plan View.

3. Place 2 × 4s that extend from the upper bridging to the floor as shown at A in the sectional view. These 2 × 4s should be nailed to the bottom edges of the bridging and to the flooring of the first floor. Find the cuts of these 2 × 4s by placing them in position and marking along the bottom edges of the bridging for the top cut.

4. Spike the 2 × 4s in position to the floor and to the bridging.

How to Build Winders on the Platform

NOTE: In building winders, a level platform is first built as in an L- or U-shaped stairway. The winder layout is made on the level platform and the winder framework is then secured to the top of this platform.

1. Lay out the locations of the winders as described.

2. Build a framework of 2 × 4s to represent the outline of the winders. Keep back from the winder layout line by the thickness of the finish riser that is to be nailed on the edge of the framework. See the dotted lines showing the outlines of the winder platforms (Fig. 23-9).

3. Measure up on the studs the height of one riser minus the thickness of the tread of the winder.

4. Temporarily nail the framework to the studs at the marks and level the framework.

5. When it is level, spike it securely to all the supporting studs.

6. If there are more winders in the platform, follow the same procedure, being sure to conform with the winder layout marked on the platform.

7. Brace the winder platforms by running cross bracing 2 × 4s where the spaces are more than 16″ in the framework of the riser platforms.

REVIEW PROBLEMS   Unit 23

1. What is the most important consideration when designing a stair?

2. Why are landings installed in a long flight of stairs?

3. State the differences between main stairs and service stairs in the average house.

4. What is the minimum headroom required for a main stair, and where is this measurement taken?

5. State the minimum width suggested for a main stair. Where is this measurement taken?

6. What is the maximum height of a stair riser and the minimum width of a stair tread?

7. Where is the best location for a straight flight of stairs?

8. Sketch a quarter-turn or L-shaped flight of stairs.

9. Is it necessary to place the platform for a half-turn or U flight of stairs midway between the two stories?

10. In addition to the safety of the people using it, what other factor should be considered when laying out the stair?

11. Upon what does the pitch of a stair depend?

12. State three rules for the sizes and proportions of treads and risers.

13. What degree of pitch is recommended for a comfortable stair?

14. State the minimum height recommended for the headroom of both the main and service stairs.

15. How much clearance should be provided at the approach and at the landings of a stair?

16. Assume that the total height from the top of one floor to the top of the next floor is 7′-9″ and that the riser is to be not over 8 1/4″ high. Find the number of risers, height of risers, the number and width of treads, and the length of the stair.

17. How is the exact height of the riser found when laying out the stair on the job?

18. Give two other names used to designate the stringer of a stair.

19. Where is the deduction made for the thickness of the tread when laying out?

20. What is meant by stair bridging?

21. Where are winders located in a flight of stairs?

22. How far from the inside edge of the internal stringer or handrail is the line of travel for winders?

23. How is the story pole used when laying out stairs?

24. How much solid stock should be left between the cutout sections for the risers and treads and the lower edge of the carriage?

25. If the stairs are to be enclosed on both sides, should the carriages or the partition studs be placed first?

26. List the tools used when framing stairs.

Part V. INSULATION

Unit 24 INSULATION

Insulation is defined as a substance which is a nonconductor. In building construction we consider both thermal (or heat) insulation and acoustical (or sound) insulation. In conjunction with the most effective use of thermal insulation we must consider vapor barriers and proper ventilation. In some cases it is the duty of the carpenter to select and install insulation, while in others the architect or the insulation contractor selects the material and describes the method of installation. In either case the carpenter should have a general knowledge of insulating methods as well as of the fundamental principles of insulation. More detailed information may be found in reference manuals or obtained from manufacturers of insulation.

FUNCTIONS OF INSULATION

Thermal insulation is used in a building to retard the transfer of heat from the warm side to the cold. Heat always moves toward a cooler area. Therefore, thermal insulation should be used where artificial heat is required or where the heat from the sun is objectionable. In the winter it prevents the excessive loss of heat from the building to the outside air. In the summer this same insulation will resist the rapid transfer of heat from the outside to the interior of the building. Most insulating materials depend upon the multitude of tiny air pockets which they contain for their effectiveness. Heat travels with difficulty through these air pockets.

Acoustical insulation is used in a building to resist the passage of sound waves either from the outside or from room areas where the noise transmission would be objectionable. Sound travels much faster through a solid substance than through air. Acoustical insulation like most thermal insulation depends upon millions of tiny air pockets. These pockets as well as an irregular surface area absorb much of the sound.

In modern construction, where the factors of economy and comfort are always to be reckoned with, we welcome insulation in a building because:

1. It prevents rapid changes in the temperature of a building in both summer and winter, thus making the building more comfortable.

2. It reduces fuel costs to a considerable extent by retaining the heat in the building. It is also often possible to use a smaller heating plant in an insulated building than in an uninsulated one.

3. It retards the spread of a fire in a building. Some insulation will not burn under any conditions, while other types burn very slowly, even when subjected to intense heat.

4. It tends to prevent excessive condensation which might cause discoloring of wall surfaces and structural damage.

5. It reduces the amount of noise transmitted from outside sources, such as airplanes, automobiles, and recreational activities to the inside of the building.

6. It reduces the sounds from the more active areas of the house, such as shops and game rooms.

7. It tends to prevent, by absorption, the possibility of disturbing echoes.

209

DESCRIPTION OF INSULATION

The materials used for building insulation are generally classified into four main divisions: mineral, vegetable or natural, plastic, and metallic.

Mineral insulation is made from rock, slag, and glass. Rock and slag wools are made fundamentally by grinding and melting and then blowing the material into a fine, wooly mass by the use of high-pressure steam. Glass wool is another type of mineral wool of fine, fluffy glass fibers. Vermiculite is a form of mica which when heated expands or explodes into granules about fifteen times their original size.

Vegetable or natural insulation is made from processed wood, sugar cane, corn stalks, and certain grasses. Cotton in the form of blankets treated to resist fire is a natural insulation. Under this division we also have cork, redwood bark, sawdust, and shavings. These may be used in their natural state or processed into various shapes and forms.

Plastic insulation is made from polystyrene and rubber. Polystyrene is an inexpensive chemical plastic that is foamed to lightweight rigid shapes. It will not crumble, rot, or disintegrate. Rubber insulation is made from synthetic rubber containing cells filled with nitrogen.

Metallic insulation is made of very thin tin plate or copper or aluminum sheets. It is also made of aluminum foil on the surface of rigid fiberboard or plasterboard. A common form is aluminum foil mounted on asphalt-impregnated kraft paper, the strength of which may be increased by the use of jute netting.

TYPES OF INSULATION

Insulation can be obtained in four general types or forms.

The loose-fill type comes in bales or bags and may be poured or blown into spaces to be insulated. It is used to fill spaces between studs or ceiling joists after the finish material is in place. It is excellent for packing into irregular spaces and for insulating walls of existing homes. This type is made from mineral wools, vermiculite, loose cork, redwood bark, and wood fibers. It serves as both a thermal and an acoustical insulator.

Blankets and batts are a flexible, fibrous type of insulation. The blankets usually come in rolls, while the batts most often come in lengths of twenty-four to forty-eight inches. Both types are available in assorted widths and thicknesses to fit between studs, joists, and rafters of different standard spacings. They may be obtained with vapor-barrier backing, which provides a flange for attachment to the framing members. Some blankets and batts are completely enclosed on all sides, with a vapor barrier on one side and also a flange. This flexible type of insulation is usually made from mineral wools or vegetable fibers. It serves as both a thermal and an acoustical insulator.

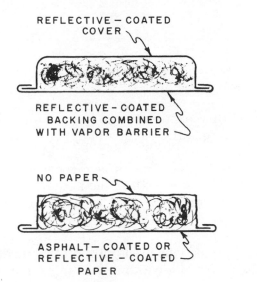

REFLECTIVE – COATED
COVER

REFLECTIVE – COATED
BACKING COMBINED
WITH VAPOR BARRIER

NO PAPER

ASPHALT – COATED OR
REFLECTIVE – COATED
PAPER

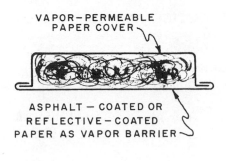

VAPOR – PERMEABLE
PAPER COVER

ASPHALT – COATED OR
REFLECTIVE – COATED
PAPER AS VAPOR BARRIER

NO PAPER BACKING OR
COVER

FIG. 24 – I   BATT OR BLANKET-TYPE
INSULATION

Insulating boards or slabs can be obtained in a great variety of sizes from ceiling or wall tiles 8″ square to sheets 4′ wide and over 10′ in length. This form is manufactured in various thicknesses from 1/2″ upward, the choice being

BOARD INSULATION

dependent upon the use and the amount of insulation required. Insulating boards or slabs are used structurally as sheathing materials and roof boards and decoratively as ceiling and wall finishes as well as for thermal and acoustical insulation. They are made from mineral wools, natural and vegetable fibers, and plastics. The structural uses of board insulation have been considered in the unit on sheathing. Board insulation as used for decorative purposes in conjunction with thermal and acoustical properties is covered in the text "Interior and Exterior Trim". The uneven surface area of the finish boards tends to absorb sound transmitted through the air.

Reflective or metallic insulation is in a class by itself. This type forms barriers of air spaces and possesses the insulating value of reflecting radiant energy. It depends upon its bright reflecting surfaces to turn back the heat. The shiny side must be installed so that it does not touch any other surface.

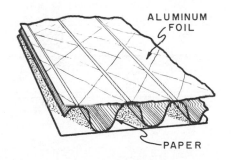

ALUMINUM
FOIL

PAPER

FIG. 24 – 2   METALLIC INSULATION

The effectiveness of metallic insulation is practically independent of the thickness of the metal itself. Some blankets, batts and boards are manufactured with a reflective insulator acting as a vapor barrier backing and flange. In some cases two sheets are combined to form a structure containing dead air cells. Metallic insulation is a thermal insulator and vapor barrier combined.

ACOUSTICS AND SOUND INSULATION

Since the average carpenter is occasionally called upon to construct sound-insulated projects such as television rooms, game or activity rooms, or basement workshop areas, some basic data and methods are presented here.

For more detailed information and treatment of acoustics the carpenter should refer to manufacturer's technical manuals and consult qualified acoustical engineers.

There are two classes of sound to be considered. The first is reverberated or reflected sounds. These are airborne sounds which continue after the actual cause has ceased and are due to reflection of sound waves from floors, walls, and ceiling. An echo would be an example of this class of sound.

There are also impact sounds that are carried by the vibrations of the structural materials themselves. Footsteps heard through the floors of a building are an example of this classification.

In many cases the use of perforated ceiling tile properly installed according to manufacturer's specifications will sufficiently reduce the sound level. These tiles are usually made from plaster, wood, or glass fiberboard with holes drilled almost through the tile. The reflected sounds which strike these units are trapped in the holes and absorbed by the soft fibrous material. Material of this type can also be installed on the upper portion of the walls to reduce the noise level further. (See the unit in "Interior and Exterior Trim" for application methods.)

Another method of soundproofing the walls of a room in frame construction is the double-stud method. To keep the thickness of the wall to a minimum, the studs may be staggered but a more efficient method is to use double plates. In any case the studs should not be in contact with each other.

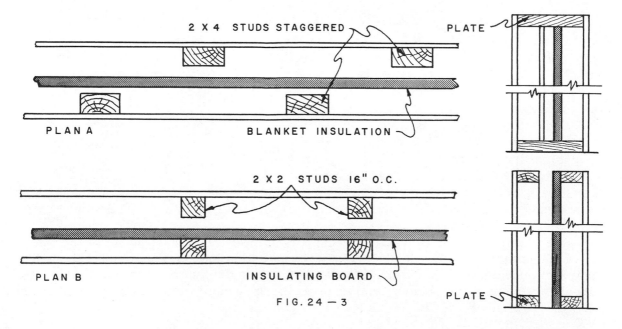

FIG. 24 — 3

Where it is desirable to soundproof a room on the first or second floor, it may be necessary to construct the floor and ceiling as shown in the illustration. A suspended acoustical tile ceiling will reduce the reverberated as well as the impact sounds. The joists or members of this suspended ceiling should have as little contact with one another as is structurally practical. A floor constructed as shown absorbs impact sounds before they reach the structural members.

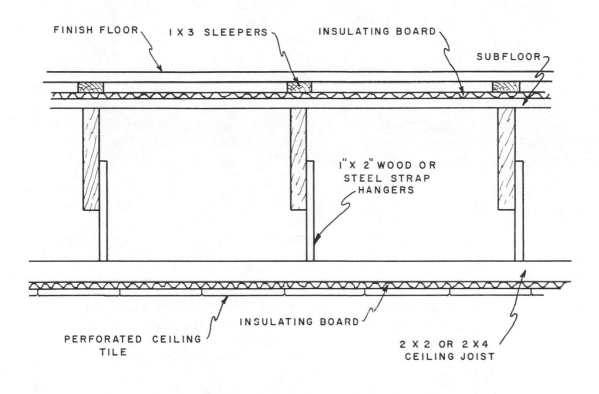

FIG. 24—4 INSULATED FLOOR AND SUSPENDED CEILING

The air spaces around doors and windows are sources of sound entrance. For this reason they should be weatherstripped, preferably by felt or rubber strips. Double glazing of windows will further reduce sound transmission. Increasing the air space will increase insulating efficiency.

Care should be taken not to install too much acoustical material, which can cause music and voices to sound muffled. Furniture, rugs, and heavy drapes will also absorb much sound. Clothes closets appropriately placed in a house or building will act as a sound barrier.

The acoustical problems involved in a meeting hall or music room are much more complicated, since sounds here have to be controlled to a much finer degree. This problem is beyond the field of the average carpenter and will not be considered here.

VAPOR BARRIERS

A vapor barrier is any material used to prevent interior water vapor from passing through the insulation and then, under certain atmospheric conditions, condensing inside exterior walls, ceiling and floor spaces. Vapor barriers are always placed on the warm side of the walls.

With the use of thermal insulation in our tightly constructed buildings of today, we are able to conserve heat, but we create a problem with the water vapor and warm air that pass through the insulation. This warm moisture condenses inside the cold exterior walls. This can cause exterior paint to peel and the wood structural members to rot. Many of our blanket and batt insulations are manufactured with a vapor barrier on one side that also serves as a flange to facilitate the placement of the insulation between structural members. Metallic or reflective insulation is in itself an efficient vapor barrier. Some wall boards are made with a metallic foil or other type of vapor-barrier backing. Foamed styrene is an excellent water-resistant insulation; it also acts as a vapor barrier.

If the insulation used does not have a vapor barrier, one should be fastened to the studs or joists before the surface finish materials are installed. A vapor barrier is always placed on the warm side of the wall or ceiling. Polyethylene film is probably the best plastic vapor barrier today. Many building papers and roofing felts are not effective vapor barriers, and their qualities as such should be investigated before using them.

In existing construction, where loose-fill insulation is poured into wall and ceiling cavities, most paints and surface coatings will help prevent the passage of water vapor. Two coats of aluminum paint, which can be covered by decorative finishes are very effective but should be used only if a vapor barrier was not installed during construction.

VENTILATION

Ventilation is a process of changing the air in a room by either natural or artificial means.

We have, with the addition of a vapor barrier to our insulated building, restricted the moist air to the inside of the structure. We must remove this moisture by ventilation or other mechanical means. This can be accomplished by the use of an exhaust fan or ventilator, which expels the vapor-laden air into the outside atmosphere. These are often placed near the source of moisture, i.e., the kitchen, bath, or laundry room. The carpenter is also often called upon to cut through the roof or walls and install louvers or ventilators. Some homes use a chemical or mechanical dehumidifier that removes the moisture from the air. Usually in the winter season, when artificial heat is used, there is little or no problem of too much vapor in the air.

Proper ventilation under roof areas prevents the accumulation of hot air in the summer, thus aiding the insulating ceiling to maintain a cooler interior temperature. This is accomplished by installing screened louvers at the highest practical location on the roof or in the gable ends of the house.

TOOLS AND EQUIPMENT FOR INSTALLING INSULATION

Hammer	Tin snips
Hook knife	Stapler
Rule	

How to Install Loose-Fill Insulation

In most cases, the manufacturer furnishes instructions for installing each particular kind of insulation. These directions should be carefully followed. Loose fill insulation is generally blown into the spaces between the studs, joists, or rafters after the lath and plaster are in place, or it may be placed by hand between the ceiling joists in an attic space. Blown-fill insulation work is generally performed by an insulation contractor and should not be attempted by any person unless he is properly safeguarded against the inhalation of particles of insulating materials.

How to Install Blanket Insulation

NOTE: Be sure the right width blanket is used to fill the spaces between the joists, studs or rafters.

1. Place the blanket between the studs. The moistureproof side of the blanket should face the inside of the room.

2. Staple the flanges of the blanket to the studs at intervals specified by the maker of the material. Blankets should be fastened between the studs so as to allow a space on both sides. When necessary, cut the blankets with a sharp knife. Short strokes are the most successful.

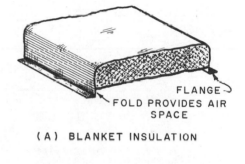

(A) BLANKET INSULATION

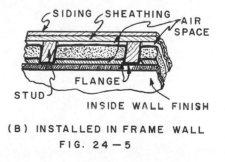

(B) INSTALLED IN FRAME WALL
FIG. 24-5

How to Apply Batt Insulation

1. Apply batts between studs. In most cases, they merely need to be pushed into place.

2. If it is necessary to fit the insulation around pipes or electrical fixtures or into odd-shaped spaces, cut it to shape with a knife.

NOTE: Sometimes the batts are attached to waterproofed paper and have flanges which are to be fastened to the studs or rafters.

3. If moistureproof paper is not attached to the batts, staple sheets of poly-ethylene film to the inside surfaces of the studs, joists, or rafters. This film should cover the entire surface of the insulated area.

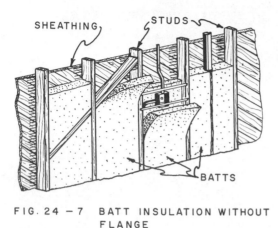

FIG. 24 — 7   BATT INSULATION WITHOUT FLANGE

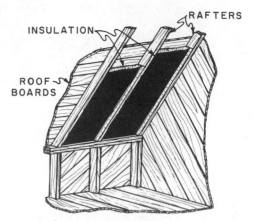

FIG. 24 — 8   BATT INSULATION WITH FLANGE

How to Apply Metallic Insulation

1. Roll insulation out on the floor to the length of joist or bay, plus 4″ to allow 2″ on each end.

2. Tear or cut off. One way is to fold it back on itself and tear along the fold.

3. Start from the end of the room on the applicator's right and staple one flange to the face of the joist or stud every 6″.

4. When the entire first flange is stapled, grasp hanging second flange with hands stretched 3′ apart and place it on the opposite joist or stud.

5. Staple every 6″

NOTE: The automatic tacker or stapling hammer may be used for this application.

How to Apply Rigid Insulation

Rigid insulation used as sheathing should be applied as described in Unit 12, Sheathing.

REVIEW PROBLEMS   Unit 24

1. Give a simple definition of insulation.

2. Which two phases of insulation are used most in the construction of homes?

3. What is the purpose of thermal insulation?

4. What is the purpose of acoustical insulation?

5. Explain six functions of insulation in modern construction.

6. List the four main classifications of insulation and describe them.

7. Describe the four general types or forms of insulating materials.

8. List four places in the home where acoustical insulation might be desired.

9. Explain the two classes of sound transmission.

10. Sketch a double-stud method of soundproofing a room.

11. Why would a suspended ceiling reduce reverberation as well as impact sounds?

12. Which types of weatherstripping are best suited to reduce sound transmission?

13. Mention other factors which reduce sound transmission.

14. What is the purpose of a vapor barrier?

15. Where should a vapor barrier be placed?

16. Describe three types of vapor barriers.

17. Are building papers and roofing felts effective vapor barriers?

18. How could a vapor barrier be applied when fill insulation is installed in an existing house?

19. Why is ventilation an important factor in our modern houses?

20. Describe three ways in which ventilation can be accomplished.

21. How is loose fill insulation installed?

22. What precaution should be taken when installing fill insulation?

23. How is blanket insulation installed?

24. If moistureproof paper is not attached to batt insulation, what material is suggested for a vapor barrier?

25. How far apart should the staples for metallic insulation be placed?

26. List the tools necessary to apply insulation.

Index